Malverd Abijah Howe

**Retaining Walls for Earth**

Including the theory of earth-pressure as developed from the ellipse of stress, with a short treatise on foundations, illustrated with examples from practice. Third Edition

Malverd Abijah Howe

**Retaining Walls for Earth**

*Including the theory of earth-pressure as developed from the ellipse of stress, with a short treatise on foundations, illustrated with examples from practice. Third Edition*

ISBN/EAN: 9783337015992

Printed in Europe, USA, Canada, Australia, Japan

Cover: Foto ©berggeist007 / pixelio.de

More available books at **www.hansebooks.com**

# RETAINING-WALLS FOR EARTH.

INCLUDING

*THE THEORY OF EARTH-PRESSURE AS DEVELOPED FROM THE ELLIPSE OF STRESS.*

WITH

A SHORT TREATISE ON FOUNDATIONS, ILLUSTRATED WITH EXAMPLES FROM PRACTICE.

BY

MALVERD A. HOWE, C.E.,

*Professor of Civil Engineering, Rose Polytechnic Institute;
Member American Society of Civil Engineers.*

*THIRD EDITION, REVISED AND ENLARGED.*

FIRST THOUSAND.

NEW YORK:
JOHN WILEY & SONS.
LONDON: CHAPMAN & HALL, LIMITED.
1896.

Copyright, 1896,
BY
MALVERD A. HOWE.

ROBERT DRUMMOND, ELECTROTYPER AND PRINTER, NEW YORK.

# CONTENTS.

## THEORY OF EARTH-PRESSURE.

|  | PAGE |
|---|---|
| Preliminary Principles | 1 |
| Resultant of Principal Stresses. Case I | 2 |
| "   "   "   "   Case II | 3 |
| "   "   "   "   Case III | 3 |
| Earth-pressure against a Vertical Plane | 9 |
| Direction of Resultant Earth-pressure against a Vertical Plane | 11 |
| Intensity of Earth-pressure against a Vertical Plane at any given Depth | 12 |
| Average Intensity of Earth-pressure against a Vertical Plane | 13 |
| Graphical Construction for finding Thrust of Earth against any Plane | 13–15 |
| General Formula for the Thrust of Earth | 15–17 |
| "   "   "   "   " Direction of the Resultant Earth-pressure | 18 |
| Plane of Rupture | 18 |
| Reliability of Theory | 19 |
| Earth Sloping Down and Away from Wall—Special Method | 21 |

## FORMULAS FOR EARTH-PRESSURE.

### Recapitulation.

| | |
|---|---|
| General Formula | 23 |
| Surface of the Earth inclined, $\alpha = 0$ | 24 |
| Surface of the Earth Parallel to the Surface of Repose | 24 |
| Surface of the Earth Horizontal | 25 |

iii

|   | PAGE |
|---|---|
| Fluid **Pressure**.................................................... | 25 |
| Graphical Construction for determining **the Thrust of Earth** against any Plane............................................. | 25 |

## STABILITY OF TRAPEZOIDAL WALLS.

|   |   |
|---|---|
| Stability against Overturning................................. | 29 |
| "         "         Sliding........................................ | 29 |
| "         "         Crushing **of Material**........................ | 30 |
| Determination of **the Width of Base of a** Trapezoidal Wall..... | 33 |

## FORMULAS FOR TRAPEZOIDAL AND TRIANGULAR WALLS.

|   |   |
|---|---|
| General Formula for Trapezoidal Walls......................... | 34 |
| Formula for Vertical Wall ..................................... | 34 |
| "         "   a Wall with a **Vertical Back** resisting a Normal Earth-pressure............................................... | 35 |
| General Formula for Triangular Walls........................... | 36 |
| Special Cases of Triangular Walls.............................. | 36 |

## FOUNDATIONS FOR WALLS RETAINING EARTH.

|   |   |
|---|---|
| General Discussion............................................. | 37 |
| Depth of Foundations.......................................... | 38 |
| Depth of Foundation when the Intensity of the Pressure upon the Base is Uniform......................................... | 39 |
| Depth of Foundation when the Intensity of the Pressure upon the Base is **Uniformly** Varying.......................... | 39–40 |
| Depth of **Foundation when** the Earth has Different Depths on Opposite **Sides of the Wall**................................ | 41 |
| Determination of the Breadth of the Base of a Trapezoidal Foundation..................................................... | 42 |
| Abutting **Power** of **Earth**...................................... | 43 |
| Bearing Power of Earth ........................................ | 43 |

CONTENTS. v

## EXAMPLES.

| | PAGE |
|---|---|
| Examples illustrating the Application of Formulas for Earth-pressure, Depth of Foundations, etc. | 44–60 |
| Examples of Retaining-wall Profiles | 61–65 |

## FOUNDATIONS.

| | |
|---|---|
| Foundations upon Rock | 66 |
| Maximum Deviation of Resultant Pressure from the Centre of the Base of the Foundation | 67 |
| Ultimate Compressive Strengths of Stone | 68 |
| Foundations upon Earth | 68 |
| Firm Earth | 68 |
| Determination of the Breadth of a Symmetrical Trapezoidal Foundation | 73 |
| Examples | 74 |
| Unsymmetrical Distribution of Pressure upon the Base of the Foundation | 75 |
| Formula for Breadth of the Base | 76 |
| Projection of Foundation Courses, etc. | 77, 78 |
| Table of Safe Projections of Courses | 79 |
| Foundations upon Soft Earth | 78 |
| Projection of Steel or Iron Beams used in Foundations | 80 |
| Table of Safe Projections | 81 |
| Pile Foundations | 81 |
| Formula for Minimum Depth of Pile | 83 |
| Trautwine's Formula | 85 |
| Engineering News Formula | 85 |
| Screw Piles | 85 |
| Sheet Piles | 86 |

## FOUNDATIONS UNDER WATER AND DEEP FOUNDATIONS.

| | |
|---|---|
| Coffer-dams | 87 |
| Timber Cribs | 87 |
| Open Caissons | 88 |

## CONTENTS.

|  | PAGE |
|---|---|
| Cushing Cylinder Piers | 89 |
| Pneumatic Caissons | 89 |
| TYPES OF EXISTING FOUNDATIONS | 90–97 |
| REFERENCES, LIST OF | 99–102 |
| DIAGRAM I | 103 |

### TABLES.

|  | |
|---|---|
| Weights of Materials | 106 |
| Angles and Coefficients of Friction | 107 |
| Values of Functions $B$, $C$, $D$, and $E$ | 108–110 |
| Natural Sines, Cosines, Tangents, and Cotangents | 111–132 |

# PREFACE TO THE SECOND EDITION.

The first edition of this work was based upon the theory advanced by Prof. Weyrauch in 1878, but owing to the length of the demonstrations used by him, it was thought advisable to present different and shorter demonstrations in this edition. To show that the new demonstrations give identical results with those obtained by Prof. Weyrauch, his demonstrations have been given in an appendix as they appeared in the first edition.

The new demonstrations are based upon the theory first advanced by Prof. Rankine in 1858. Those readers who are familiar with Rankine's Ellipse of Stress can omit pages 1 to 9, inclusive, in following the demonstrations.

An attempt has been made to present the theory in a shape easily followed by those who have only a knowledge of algebra, geometry, and trigonometry; whenever calculus has been resorted to, the work has been simplified as much as possible. For convenience in practice, the formulas have been arranged in a condensed shape in Part I, and are followed by numerous examples illustrating their application.

The values of various coefficients have been computed and tabulated and will be found to very materially decrease the labor of substitution in the formulas.

It is hoped that the introduction of a brief treatment of the supporting power of earth in the case of foundations, as well as the formula for determining the breadth of the base of a retaining-wall, will prove acceptable.

For valuable help in the verification of proofs of formulas, and the critical reading of the whole text, I acknowledge the kind assistance of Prof. Thos. Gray.

M. A. H.

## PREFACE TO THE THIRD EDITION.

In this edition a large number of examples illustrating several profiles of retaining-walls and types of foundations selected from existing structures have been included. The Appendix of the second edition has been replaced by a treatise on Foundations sufficiently short and, the author believes, sufficiently complete for the use of technical schools, if judiciously supplemented by lectures or references to descriptions of existing structures.

M. A. H.

Terre Haute, Ind., Nov. 1896.

# NOMENCLATURE.

$\phi$ = the angle of repose, or the maximum angle which any force acting upon any plane within the mass of earth can make with the normal to the plane.

$\epsilon$ = the angle made by the surface of the earth with the horizontal; $\epsilon$ is *positive* when measured *above* and *negative* when measured *below* the horizontal.

$\alpha$ = the angle which the back of the wall makes with the vertical passing through the heel of the wall; $\alpha$ is *positive* when measured on the *left* and *negative* when measured on the *right* of the vertical.

$\delta$ = the angle which the direction of the resultant earth-pressure makes with the horizontal.

$\phi'$ = the angle of friction between the wall and its foundation.

$\phi''$ = the angle of friction between the back of the wall and the earth.

$H$ = the vertical height of the wall in feet.

$h$ = the depth of earth in feet which is equivalent to a given load placed upon the surface of the earth.

$B'$ = the width in feet of the top of the wall.

$B$ = the width in feet of the base of the wall.

$Q$ = the distance in feet from the toe of the wall to the point where $R$ cuts the base.

# NOMENCLATURE.

$P$ = the resultant earth-pressure in pounds against a vertical wall.

$E$ = the resultant earth-pressure in pounds against any wall.

$R$ = the resultant pressure in pounds on the base of the wall.

$G$ = the total weight in pounds of material in the wall.

$\gamma$ = the weight in pounds of a cubic foot of earth.

$W$ = the weight in pounds of a cubic foot of wall.

$p$ = the intensity of the pressure in pounds on the base of the wall at the toe.

$p'$ = the intensity of the pressure in pounds on the base of the wall at the heel.

$p_0$ = the average intensity of the pressure in pounds on the base of the wall.

$x = H \tan \alpha$.

$x''$ and $x'$ = depth of the base of the foundation below the earth surface.

$B''$ = breadth of the base of the foundation.

$o$ = the offset of a foundation course.

$G'$ = the total weight of the material above the base of the foundation.

# THEORY OF EARTH-PRESSURE.

*Preliminary Principles.*—Before demonstrating the general formula for the thrust of earth against a wall, it will be necessary to establish the relations between the stresses in an unconfined and homogeneous granular mass.

\* In Fig. 1 let $ABC$ be any small prism within a granu-

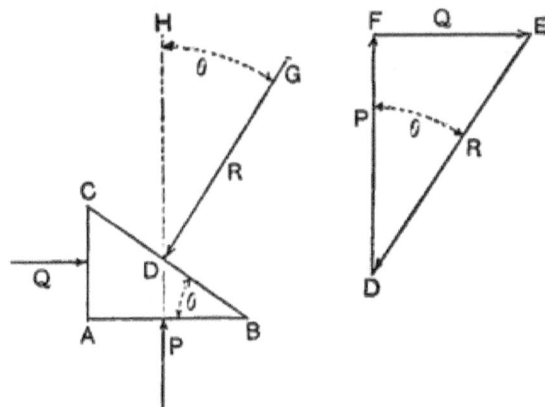

Fig. 1.

lar mass which is in equilibrium un er the action of the three stresses $P$, $Q$, and $R$, having the intensities $p$, $q$, and $r$ respectively.

---

\* In all the demonstrations which follow, the dimension perpendicular to the page will be considered as unity.

Let $\theta$ represent the angle of inclination of the plane $CB$ with $AB$, and the angle at $A$ be a right angle.

The planes $AB$ and $AC$ are called planes of principal stress, and $P$ and $Q$ are called principal stresses.

**CASE I.** *If the principal stresses are of the same kind and their intensities the same,* **then will the resultant stress** *on any third plane be normal to that plane and its intensity be equal to that* of *either principal* **stress.**

In Fig. 1, for convenience, let $AB = 1$, then $AC = \tan \theta$, and $CB = \dfrac{1}{\cos \theta}$. Hence

$$P = p, \quad Q = q \tan \theta = p \tan \theta, \text{ since } p = q, \text{ and } R = \dfrac{r}{\cos \theta}.$$

Since $P$, $Q$, and $R$ are in equilibrium, they will form a closed triangle, as shown on the right in Fig. 1. Hence

$$R^2 = P^2 + Q^2,$$

or

$$\dfrac{r^2}{\cos^2 \theta} = p^2 + p^2 \tan^2 \theta = p^2(1 + \tan^2 \theta);$$

$$\therefore r = p = q.$$

Also, $\qquad R \cos FDE = P,$

or $\qquad \dfrac{r}{\cos \theta} \cos FDE = p; \quad \text{but } r = p.$

Hence $\qquad \cos \theta = \cos FDE = \cos HDG;$

$\therefore HDG = \theta \quad \text{and} \quad R \text{ is normal to } CB.$

CASE II. *If the principal stresses are not of the same kind but their intensities the same, then will the resultant make the angle θ with the direction of the principal stress, but on the opposite side from that on which the resultant in Case* I *lies, and its intensity be equal to that of either principal stress.*

The demonstration of Case I proves this principle if Fig. 1 is replaced by Fig. 2.

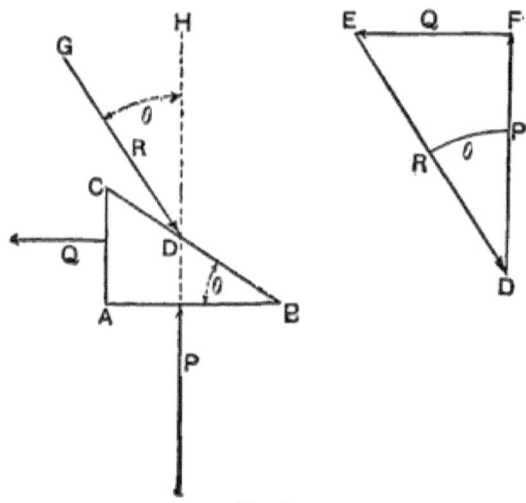

FIG. 2.

CASE III. *Given the principal stresses of the same kind but having unequal intensities, to determine the intensity and direction of the resultant stress on any third plane.*

Let $P$ and $Q$ be compressive and the intensity $p >$ the intensity $q$.

The following identities can be written:

$$p = \tfrac{1}{2}(p+q) + \tfrac{1}{2}(p-q),$$

and

$$q = \tfrac{1}{2}(p+q) - \tfrac{1}{2}(p-q);$$

or the resultant intensity on the plane $CB$ may be considered as being the resultant of two intensities, one being the intensity of the resultant stress caused by two like principal stresses having the same intensity $\frac{1}{2}(p+q)$, and the other the intensity of the resultant stress caused by two unlike principal stresses having the same intensity $\frac{1}{2}(p-q)$.

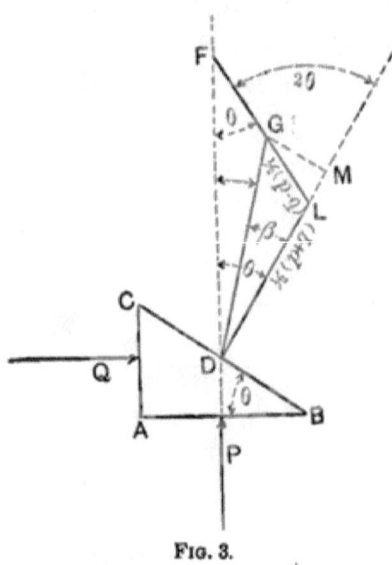

Fig. 3.

The intensity of the resultant stress caused by the first two principal stresses will be, by Case I, $\frac{1}{2}(p+q)$, and the direction of the resultant will be normal to the plane $CB$. By Case II the resultant of the second pair of principal stresses will make the angle $\theta$ with the direction of $P$, and its intensity will be $\frac{1}{2}(p-q)$; then the resultant intensity can be found as follows:

In Fig. 3 draw $MD$ normal to $BC$, and make $LD = \frac{1}{2}(p+q)$; with $L$ as a centre and $LD$ as radius, describe an arc cutting $FD$ at $F$. Then the angle $LFD = LDF = \theta$. Lay off $LG = \frac{1}{2}(p-q)$, and draw $GD$, which is the resultant intensity, and the intensity of the resultant stress on $CD$ caused by the two principal stresses $P$ and $Q$. $GD$ also represents the direction of the resultant stress $R$.

Since the intensities of the principal stresses remain constant, $\frac{1}{2}(p+q)$ and $\frac{1}{2}(p-q)$ will remain the same for any inclination of the plane $CB$; hence the intensity $r$ of the resultant depends upon the angle $\theta$ when $p$ and $q$ are given.

From **Fig. 3**,

$$GL \cos 2\theta = LM \quad \text{and} \quad GL \sin 2\theta = GM,$$

$$DM = DL + LM = \tfrac{1}{2}(p+q) + \tfrac{1}{2}(p-q)\cos 2\theta,$$

$$\overline{GD}^2 = r^2 = \overline{GM}^2 + \overline{DM}^2,$$

or

$$r = \sqrt{p^2 \cos^2\theta + q^2 \sin^2\theta}, \quad \ldots \quad (a)$$

which is the general expression for the intensity of the resultant stress of a pair of principal stresses.

As the angle $\theta$ changes, the angle $\beta$ will also change, and it will have its maximum value when the angle $LGD = 90°$. This is easily proven as follows:

With $L$ as centre and $GL$ as radius describe an arc; then $\beta$ will have its maximum value when the line $DG$ is tangent to the arc; but when $DG$ is tangent to the arc the angle $LGD$ is a right angle, since $LG$ is the radius of the arc.

$$\sin \max \beta = \frac{p-q}{p+q}, \quad \ldots \quad (b)$$

from which the following can be easily obtained:

$$\frac{p}{q} = \frac{1 + \sin \max \beta}{1 - \sin \max \beta}, \quad \ldots \quad (c)$$

which expresses the limiting ratio of the intensities of the principal stresses consistent with equilibrium, $p$ being greater than $q$.

**Case IV.** *Given the intensity and direction of the resultant stress on any plane,* **and the** *value of max $\beta$, to determine the intensities and* **directions** *of the principal stresses.*

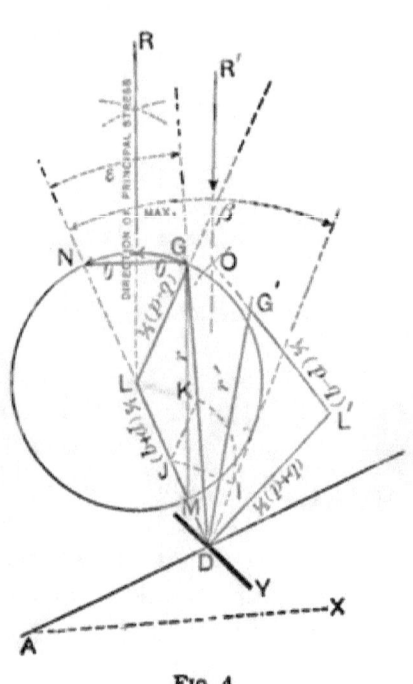

Fig. 4.

Let $AD$ represent the given plane and $GD$ the direction and intensity of the resultant stress at the point $D$.

Draw $DL$ normal to $AD$, and draw $DI$, making the angle max $\beta$ with $LD$. At any point $J$ in $DL$ describe an arc tangent to $DI$, cutting $GD$ in $K$ and draw $GL$ parallel to $KJ$; with $L$ as a centre and $LG$ as radius describe a circumference. This circumference will pass through $G$ and be tangent to $DI$; hence $\dfrac{GL}{DL} = \sin \max \beta$.

Since $\sin \max \beta = \dfrac{p-q}{p+q}$, and $GL$ and $LD$ are components of $r$,

$$GL = \tfrac{1}{2}(p-q) \quad \text{and} \quad DL = \tfrac{1}{2}(p+q);$$

then $ND = NL + LD = \tfrac{1}{2}(p-q) + \tfrac{1}{2}(p+q) = p,$

and $MD = LD - LM = \tfrac{1}{2}(p+q) - \tfrac{1}{2}(p-q) = q,$

which completely determines the intensities of the principal stresses.

According to Case III, the direction of the greater principal stress bisects the angle between the prolongation of $LM$ and the line $GL$; hence $RL$ represents the direction of the greater principal stress, and that of the other is at right angles to $RL$.

The above intensities and directions being determined, the intensity of the resultant stress on any other plane passing through $D$ is easily determined as follows:

Let $DY$ represent any plane passing through $D$, draw $DL'$ normal to $DY$ and equal to $\tfrac{1}{2}(p+q)$. Draw $R'D$ parallel to $RL$, and with $L'$ as a centre and $L'D$ as radius describe an arc cutting $R'D$ at $O$, and make $L'G' = \tfrac{1}{2}(p-q)$; then $G'D = r' =$ the intensity of the resultant stress on $DY$.

It is clear that if the value of max $\beta$ can be obtained for a mass of earth that the construction of Fig. 3 can be employed in determining the intensity of the earth-pressure at any point in *any plane* within the mass.

It has been established by experiment that if a body be placed upon a plane, that (as the plane is made to incline to the horizontal) at some angle of inclination the body will commence to slide down the plane, and that this angle depends largely upon the *character* of the surfaces in contact.

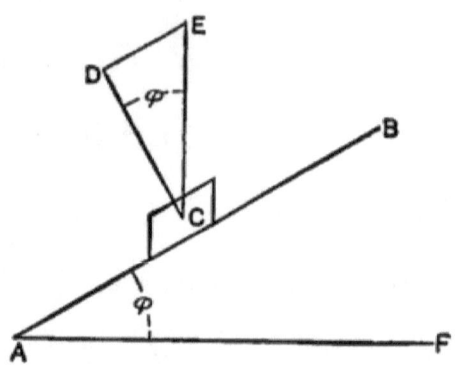

Fig. 5.

In Fig. 5 let $AB$ represent a plane inclined at the angle $\phi$ with the horizontal, and $C$ any mass just on the point of sliding down the plane. Let $EC$ represent the weight of the mass $C$, and $ED$ and $DC$ the components respectively parallel and normal to the plane $AB$. Then $DE$ is the force required to just keep the mass $C$ from sliding down the plane, assuming the plane to be perfectly smooth, or if the plane is rough this force represents the effect of friction.

$$\frac{DE}{DC} = \tan \phi,$$

or when the mass $C$ is about to slide, the resultant pressure $EC$ on $AB$ makes the angle $\phi$ with the normal to the plane, the angle $\phi$ being the inclination of the plane $AB$, and is called the angle of friction.

In the case of earth, considered as a dry granular mass, the inclination of the steepest plane upon which earth will not slide is called the angle of repose, and the plane the surface of repose.

From the above, then, it follows that in a mass of earth the resultant pressure on any plane cannot make an angle with the normal to that plane which is greater than the angle of repose $\phi$; therefore the construction of Case IV applies to earth when max $\beta$ is replaced by $\phi$. The values of $\phi$ for earth under various conditions are given in Table II.

The preceding principles will now be applied in determining the thrust of earth against a retaining-wall.

## EARTH-PRESSURE.

In order that the formulas may not become too complex for practical use, it will be assumed that the earth is a homogeneous granular mass without cohesion. The surface of the earth will be considered to be a plane, and the length of the mass measured normally to the page as unity.

*Given the intensity and direction of the resultant stress at any point in any plane parallel to the surface of the earth, the inclination of the surface of the earth with the horizontal, and the angle of repose, to determine the intensity and direction of the resultant stress on a vertical plane passing through the same point.*

---

*For comparison, see the "Technic," 1888; a construction by Prof. Greene.

The construction follows (see Fig. 4, above) directly from Rankine's Ellipse of Stress.

In Fig. 6 let $BQ$ represent the surface of the earth, and $D$ any point in the plane $AD$ parallel to $BQ$; draw $DQ$ normal to $AD$, and make the vertical $GD$ equal to $QD$; then $GD \cdot \gamma$ is the intensity of the resultant pressure at $D$. Draw $DM$, making the angle $\phi$ with $LD$, and with $L$ as centre describe an arc tangent to $DM$ and passing through $G$; then by Case IV $LG \cdot \gamma = \frac{1}{2}(p-q)$, $LD \cdot \gamma = \frac{1}{2}(p+q)$,

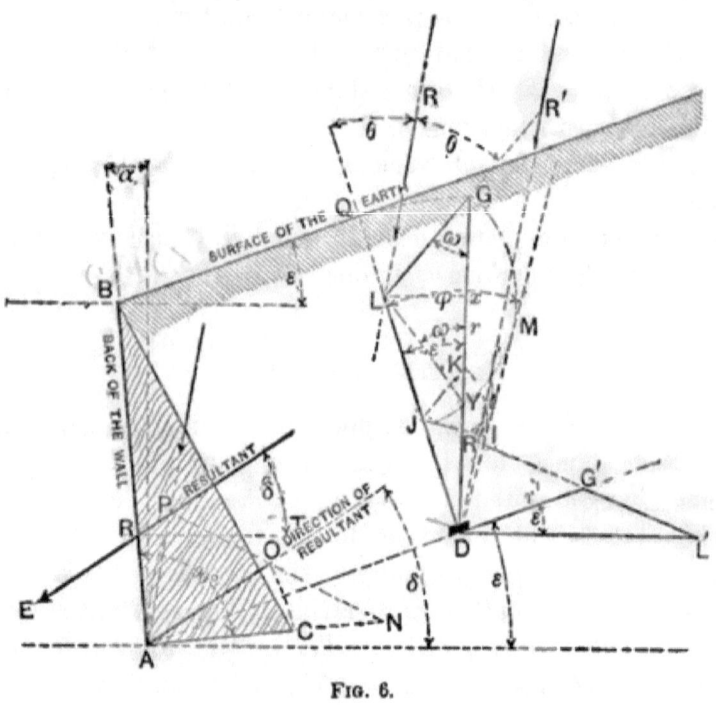

FIG. 6.

and $RL$ bisecting the angle $QLG$ is the direction of the greater principal stress. To determine the intensity and direction of the resultant stress at $D$ on a vertical plane, proceed according to Case IV. Draw $R'D$ parallel to $RL$ and $DL' = DL$ normal to $DG$. With $L'$ as a centre and $L'D$ as radius describe an arc cutting $R'D$ at $R''$, and make

$L'G' = LG$; then $DG'$ represents the direction of the resultant stress, and $DG' \cdot \gamma$ the intensity of the resultant.

In Fig. 6 the angle $R'DL' = DR''L' = 90° - \omega + \theta$. $\therefore G'L'D = 2\omega - 2\theta$. But $2\theta = \omega + \epsilon$; hence $G'L'D = \omega - \epsilon$.

Draw $LY = LG$; then the angle $DLY = \omega - \epsilon$. $\therefore$ Since $LD = DL'$ and $LY = LG = L'G'$, the triangle $G'L'D$ equals the triangle $LYD$ and the angle $G'DL' = \epsilon$; or *the direction of the resultant earth-pressure against a vertical plane is parallel to the surface of the earth.*

From Fig. 6,

$$\tfrac{1}{2}(p - q) \cos \omega = GX \cdot \gamma,$$
$$\tfrac{1}{2}(p - q) \sin \omega = LX \cdot \gamma,$$
$$\tfrac{1}{2}(p + q) \cos \epsilon = DX \cdot \gamma.$$

Now $\quad DY = DG' = DG - 2GX,$

or
$$DG' \cdot \gamma = DG \cdot \gamma - (p - q) \cos \omega$$
$$= \tfrac{1}{2}(p + q) \cos \epsilon - \tfrac{1}{2}(p - q) \cos \omega,$$
$$\tfrac{1}{2}(p + q) : \sin \omega \; :: \; \tfrac{1}{2}(p - q) : \sin \epsilon,$$

and
$$\sin \omega = \frac{p + q}{p - q} \sin \epsilon,$$

or
$$\cos \omega = \sqrt{1 - \left(\frac{p+q}{p-q}\right)^2 \sin^2 \epsilon} = \sqrt{\frac{(p-q)^2 - (p+q)^2 \sin^2 \epsilon}{(p-q)^2}},$$

and since $\quad \tfrac{1}{2}(p + q) \sin \phi = \tfrac{1}{2}(p - q),$

$$\cos \omega = \frac{1}{\sin \phi} \sqrt{\cos^2 \epsilon - \cos^2 \phi}.$$

Substituting this value for cos ω in the equation for $DG' \cdot \gamma$, it becomes

$$DG' \cdot \gamma = \tfrac{1}{2}(p+q)\cos\epsilon - \tfrac{1}{2}(p-q)\frac{1}{\sin\phi}\sqrt{\cos^2\epsilon - \cos^2\phi},$$

or since
$$\frac{1}{\sin\phi} = \frac{p+q}{p-q},$$

$$DG' \cdot \gamma = \tfrac{1}{2}(p+q)\{\cos\epsilon - \sqrt{\cos^2\epsilon - \cos^2\phi}\}.$$

In a similar manner,

$$DG \cdot \gamma = \tfrac{1}{2}(p+q)\{\cos\epsilon + \sqrt{\cos^2\epsilon - \cos^2\phi}\},$$

and
$$\frac{DG'}{DG} = \frac{\cos\epsilon - \sqrt{\cos^2\epsilon - \cos^2\phi}}{\cos\epsilon + \sqrt{\cos^2\epsilon - \cos^2\phi}};$$

hence
$$DG' = DG\frac{\cos\epsilon - \sqrt{\cos^2\epsilon - \cos^2\phi}}{\cos\epsilon + \sqrt{\cos^2\epsilon - \cos^2\phi}}.$$

Let $x=$ the *vertical* distance between the two planes $BQ$ and $AD$, then

$$DG = DQ = x\cos\epsilon.$$

$$\therefore DG' \cdot \gamma = (x)\gamma\cos\epsilon\frac{\cos\epsilon - \sqrt{\cos^2\epsilon - \cos^2\phi}}{\cos\epsilon + \sqrt{\cos^2\epsilon - \cos^2\phi}},$$

which is the expression for the intensity of the resultant earth-pressure on a vertical plane at any depth $x$ below the surface.

Let

$$* \, A = \cos\epsilon\frac{\cos\epsilon - \sqrt{\cos^2\epsilon - \cos^2\phi}}{\cos\epsilon + \sqrt{\cos^2\epsilon - \cos^2\phi}}. \quad . \quad . \quad (d)$$

---

* See Rankine's Applied Mechanics; Alexander's Applied Mechanics; Theories of Winkler and Mohr.

# THEORY OF EARTH-PRESSURE.

The average intensity of the resultant earth-pressure on a vertical plane of the length $x$ will be

$$\left(\frac{x}{2}\right)\gamma A,$$

and hence the total pressure will be

$$P = \frac{x^2}{2}\gamma A. \quad \ldots \quad \ldots \quad (e)$$

Since the intensities of the pressures are uniformly varying from the surface, and increasing as $x$ increases, the application of the resultant thrust will be at a depth of $\frac{2}{3}x$ below the surface.

Considering the earth as an unconfined mass, the above formula is perfectly general and can be applied under all conditions, including the case when $\epsilon$ is negative.

The resultant stress on any plane as $AB$, Fig. 6, can be found by applying the principles of Case IV. Draw $PA$ parallel to $RL$, make $AN = LD$ and $NO = LG$; then $AO$ represents the direction of the resultant pressure on $AB$. Make $AC = AO$; then the area of the triangle $ABC$ multiplied by $\gamma$ is the total pressure on the plane $AB$, and this pressure is applied at $\frac{2}{3}AB$ below $B$.

In unconfined earth this construction is perfectly general and applies to *any plane*. It also applies equally well to curved profiles. An example illustrating the application of the method will be given in the *applications*. See pages 22 and 23.

The following graphical construction, Fig. 7, is more convenient than that of Fig. 6.

As before, let $BE$ represent the surface of the earth, and

$AD$ a plane parallel to the surface. At any point $D$ in this plane, draw $DE$ vertical and make $DF = DE$; draw $FG$ horizontal and make the angle $HFD = \phi$.

With $L$ as a centre, describe an arc passing through $G$ and tangent to $MF$; then with $L$ as a centre and $LF$ as

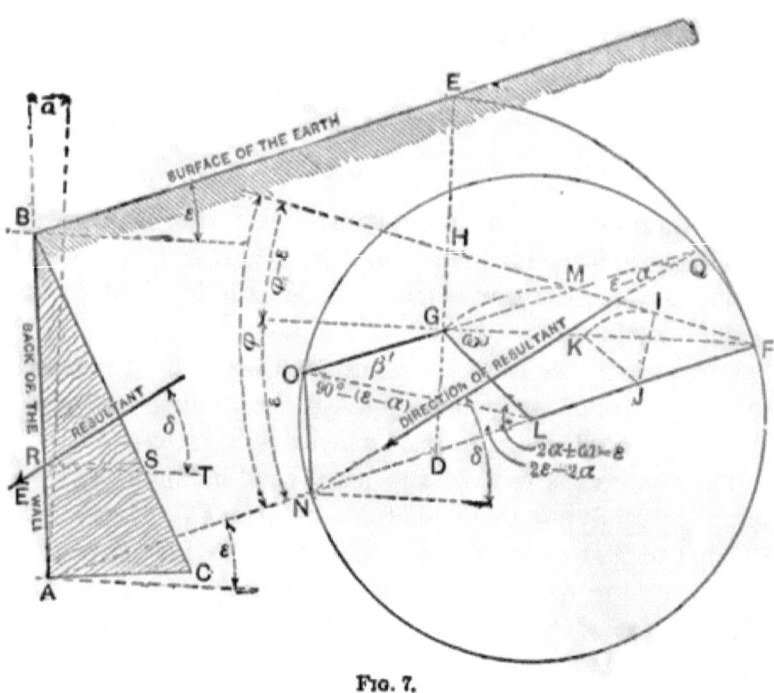

Fig. 7.

radius, describe the circumference $FON$, cutting $AD$ at $N$; through $N$ draw $NO$ parallel to $AB$, then draw $AC$ normal to $AB$ and equal to $OG$. The area of the triangle $ABC$ multiplied by $\gamma$ will be the total earth-pressure on $AB$. To determine the direction of the thrust prolong $OG$ to $Q$, then $QN$ is the direction of the thrust.

That this construction is equivalent to that of Fig. 6 is

proved as follows. The triangle $GLF$ of Fig. 7 equals the triangle $GLD$ of Fig. 6.

$$\therefore GL\cdot\gamma = \tfrac{1}{2}(p-q) \quad \text{and} \quad LF\cdot\gamma = LO\cdot\gamma = \tfrac{1}{2}(p+q).$$

In Fig. 6, the angle $NAP = NPA = 90° - \tfrac{1}{2}(\omega-\epsilon) - \alpha$.

$$\therefore ONA = \omega - \epsilon + 2\alpha.$$

In Fig. 7, the angle $OLN = 2\epsilon - 2\alpha$. But $GLN = \omega + \epsilon$.

$$\therefore GLO = \omega - \epsilon + 2\alpha,$$

and $GO$ of Fig. 7 equals $AO$ of Fig. 6.

In Fig. 7, the angle $QNO = 90° - \beta'$.
In Fig. 6, the angle $OAB = 90° - \beta'$.

Therefore the direction of the thrust is the same in both constructions.

The two constructions given above are all that is required to determine the thrust of earth upon any plane within the mass of earth, as one can be used as a check upon the other; but as a formula is often very convenient, a general formula will now be deduced which will enable one to determine the values of $E$ and $\delta$ for any plane within a mass of earth.

GENERAL FORMULA FOR THE THRUST OF EARTH.

In Fig. 8, let $BQ$ represent the surface of the earth and $AB$ any plane upon which the earth-pressure is desired.

Draw $AD$ parallel to $BQ$ and let the vertical distance $QD = FA = x$.

From (e) the earth-pressure upon $FA$ is parallel to the surface and equal to

$$P = \frac{x^2}{2}\gamma A.$$

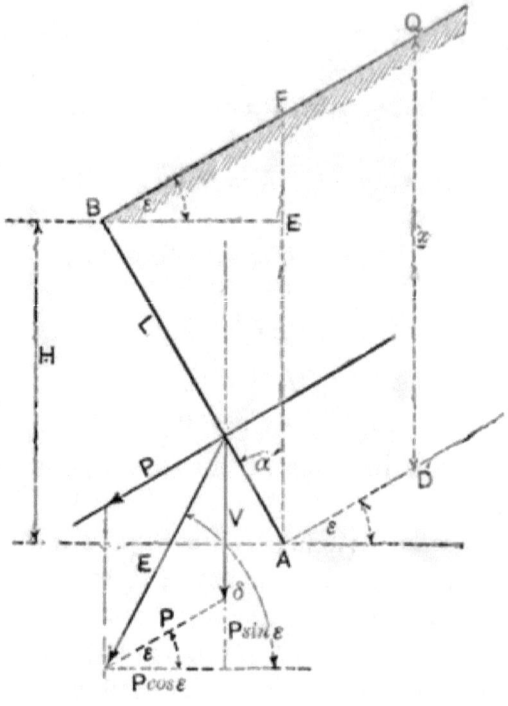

Fig. 8.

But $AF = x = H(1 + \tan \alpha \tan \epsilon) = H\dfrac{\cos(\epsilon - \alpha)}{\cos \alpha \cos \epsilon}$;

$$\therefore P = \frac{H^2 \gamma}{2} \frac{\cos^2(\epsilon - \alpha)}{\cos^2 \alpha \cos^2 \epsilon} A. \quad . \quad . \quad . \quad (f)$$

Now the thrust $P$ combined with the weight of the prism $ABF$ must produce the resultant pressure upon $AB$.

## THEORY OF EARTH-PRESSURE.

Then from Fig. 8,

$$V = \frac{H^2\gamma}{2} \tan \alpha \, (1 + \tan \alpha \tan \epsilon)$$

$$= \frac{H^2\gamma}{2} \frac{\sin \alpha \cos (\epsilon - \alpha)}{\cos^2 \alpha \cos \epsilon}, \quad . \quad (g)$$

$$E = \sqrt{(V + P \sin \epsilon)^2 + (P \cos \epsilon)^2} = \sqrt{V^2 + P^2 + 2 VP \sin \epsilon}.$$

Substituting $(f)$ and $(g)$ in this it becomes

$$E = \frac{H^2\gamma}{2} \frac{\cos (\epsilon - \alpha)}{\cos^2 \alpha \cos \epsilon} \times$$

$$\sqrt{\sin^2 \alpha + 2 \sin \alpha \sin \epsilon \cos (\epsilon - \alpha) \frac{A}{\cos \epsilon} + \cos^2(\epsilon - \alpha) \frac{A^2}{\cos^2 \epsilon}},$$

which becomes, by replacing $A$ by its value from $(d)$,

$$E = \frac{H^2\gamma}{2} \frac{\cos (\epsilon - \alpha)}{\cos^2 \alpha \cos \epsilon} \times$$

$$\sqrt{\begin{array}{l} + \sin^2 \alpha \\ + 2 \sin \alpha \sin \epsilon \cos (\epsilon - \alpha) \dfrac{\cos \epsilon - \sqrt{\cos^2 \epsilon - \cos^2 \phi}}{\cos \epsilon + \sqrt{\cos^2 \epsilon - \cos^2 \phi}} \\ + \cos^2 (\epsilon - \alpha) \left\{ \dfrac{\cos \epsilon - \sqrt{\cos^2 \epsilon - \cos^2 \phi}}{\cos \epsilon + \sqrt{\cos^2 \epsilon - \cos^2 \phi}} \right\}^2 \end{array}}, \quad . \quad (1)$$

which is the general equation for the thrust of earth upon *any plane* within the mass.

To determine the direction of the thrust of the earth, let $\delta$ be the angle which the direction of the thrust makes with the horizontal; then, from Fig. 8,

$$\tan \delta = \frac{V}{P \cos \epsilon} + \tan \epsilon.$$

Substituting the values of $V$ and $P$ given above, this becomes

$$\tan \delta = \frac{\sin \alpha \cos \epsilon + \sin \epsilon \cos (\epsilon - \alpha) A}{\cos \epsilon \cos (\epsilon - \alpha) A}, \quad (1a)$$

where

$$A = \cos \epsilon \frac{\cos \epsilon - \sqrt{\cos^2 \epsilon - \cos^2 \phi}}{\cos \epsilon + \sqrt{\cos^2 \epsilon - \cos^2 \phi}}. \quad (d)$$

Equations (1) and (1a) are readily reduced to more simple forms for special cases. These forms will be found in Part I.

*The Plane of Rupture.*—Although it is not necessary to know the position of the plane of rupture in order to determine the thrust of the earth, yet it may be of interest to know its position, which can be easily determined as follows:

The plane of rupture will be back of the wall and pass through the heel of the wall. The resultant earth-pressure will make the angle $\phi$ with the normal to this plane. Now the tangent of the angle which the direction of the resultant earth-pressure on any plane makes with the horizontal is determined from the formula

$$\tan \delta = \frac{\sin \alpha}{\cos (\epsilon - \alpha) A} + \tan \epsilon.$$

If $\omega$ represents the angle which the plane of rupture makes with the vertical passing through the heel of the wall, $\alpha = \omega$ and $\delta = \phi + \omega$.

$$\tan (\phi + \omega) = \frac{\sin \omega}{\cos (\epsilon - \omega) A} + \tan \epsilon,$$

from which the value of $\omega$ can be determined for any case.

For the case where $\epsilon = \phi$, $\epsilon$ being positive with respect to the wall and *negative with respect to the plane of rupture*, the above equation becomes

$$\tan(\phi + \omega) = \frac{\sin \omega}{\cos(\phi + \omega) \cos \phi} - \tan \phi,$$

which is satisfied when $\omega = 90° - \phi$.

For the case where $\epsilon = 0$,

$$\tan(\phi + \omega) = \frac{\sin \omega}{\cos \omega \tan^2\left(45° - \frac{\phi}{2}\right)},$$

which is satisfied when $\omega = 45° - \frac{\phi}{2}$.

*Reliability of the Preceding Theory.*—The preceding theory is based upon the assumptions that the earth is a homogeneous mass and without cohesion, and the formulas are deduced under the assumption that the surface of the earth is a plane.

All writers on the subject have considered the earth as a homogeneous mass and, with a few exceptions, without cohesion.

Old and recent experiments indicate that cohesion has very little effect upon the pressure of the earth, which explains why it has not been considered by most writers.

The assumption of a plane earth-surface is necessary whenever practical formulas and direct graphical constructions for obtaining the thrust of the earth are obtained. General formulas can be deduced for any character of surface, but they are too complex for practical use. Those graphical constructions which do not require a plane earth-surface are not direct in their solution of the problem, but require a series of trials to obtain the maximum thrust.

If the earth-surface is not a plane, one can be assumed which will give the thrust of the earth sufficiently exact for all practical purposes.

For unconfined earth no exceptions can be taken to the preceding theory, the assumptions upon which it is based being accepted, and for confined earth the theory must be true when the direction of the principal stress passing through the heel of the wall lies entirely within the earth.

For all cases in which $\alpha$ and $\epsilon$ are positive the theories of *Rankine, Winkler, Weyrauch,* and *Mohr* agree and give identical results with the preceding theory, as they should, being founded upon the same assumptions.

When $\alpha$ is negative *Weyrauch* does not consider his theory reliable, and his equations lead to indeterminate results.

*Winkler* and *Mohr* consider their theories reliable whenever the direction of the principal stress passing through the heel of the wall lies entirely within the earth.

*Rankine's* method of considering the case where $\alpha$ is negative is equivalent to assuming that the introduction of a wall does not affect the stresses within the mass.

It may be concluded that the preceding theory is perfectly exact when $\alpha$ and $\epsilon$ are positive; and when $\alpha$ or $\epsilon$ is negative that the stresses obtained will be the maximum which under any circumstances can exist.

For the case where $\epsilon$ is negative the stress obtained will be considerably larger than the actual stress (when a wall is introduced), depending upon the magnitude of $\epsilon$. For small values of $\epsilon$ the results will be practically correct. For large values of $\epsilon$ the following method can be employed in determining the thrust of the earth. The method depends upon the *assumption* that the pressure of the earth is normal to the back of the wall. This may or may not be the case, but it appears to be the most consistent assumption to make for this rare and not important case.

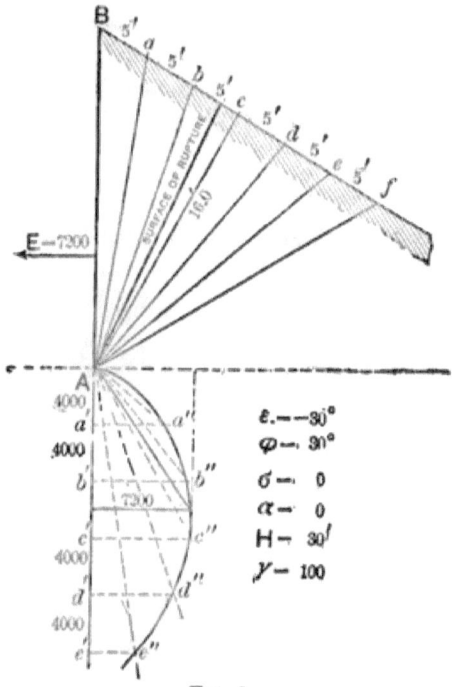

Fig. 9.

* In Fig. 9, let $AB$ be the back of the wall and $Bf$ the surface of the earth. Make $Ba = ab = bc = cd =$ etc. Some prism $BAa$ or $BAb$ or $BAc$, etc., will produce the maximum thrust on the wall; and when this maximum thrust is produced, the resultant pressure on the plane $Aa$

---

* See Van Nostrand's Magazine, XVII, 1877, p. 5. "New Constructions in Graphical Statics," by H. T. Eddy, C.E., Ph.D.

or $Ab$ or $Ac$, etc., will make the angle $\phi$ with the normal to the plane.

On the vertical line $Ad'$ lay off $Aa'=a'b'=b'c'$, etc., and draw $Aa''$ making the angle $\phi$ with the normal to $Aa$, $Ab''$ making the angle $\phi$ with the normal to $Ab$, etc.; then draw $a'a''$, $b'b''$, etc., perpendicular to $AB$, and draw a curve through $Aa''$, $b''$, $c''$, etc. Then there will be a maximum distance parallel to $a'a''$ between $Ad'$ and this curve which will be proportional to the thrust of the earth against $AB$. This maximum distance multiplied by the altitude $Ac \div 2$ and the product by $\gamma$, the weight of a cubic foot of earth, will be the pressure of the earth.

This method is perfectly general and can be applied in any case.

If the earth-pressure is assumed to have the direction given by the formulas of the preceding theory, the construction will give the same value of $E$, the pressure of the earth.

Some writers assume that the direction of $E$ makes the angle $\phi'' = \phi$ with the normal to the back of the wall in all cases. This assumption cannot be correct until the wall commences to tip forward, and then it is doubtful that such is the case unless the earth and wall are perfectly dry.

To be on the side of safety in every case, it is better to take the direction of $E$ as given by the above theory.

The construction of Fig. 9 will give the maximum thrust for any assumed direction for any case.

# FORMULAS FOR EARTH-PRESSURE.

In the following formulas $\alpha$ and $\epsilon$ are considered as positive, and the wall is assumed to be one foot long.

CASE I. *General case of inclined earth-surface and inclined back of wall.*

$$E = \frac{H^2 \gamma}{2} \frac{\cos(\epsilon - \alpha)}{\cos^2 \alpha \cos \epsilon} \times$$

$$\sqrt{\sin^2 \alpha + \cos^2(\epsilon - \alpha) \left\{ \frac{\cos \epsilon - \sqrt{\cos^2 \epsilon - \cos^2 \phi}}{\cos \epsilon + \sqrt{\cos^2 \epsilon - \cos^2 \phi}} \right\}^2 + 2 \sin \epsilon \sin \alpha \cos(\epsilon - \alpha) \left\{ \frac{\cos \epsilon - \sqrt{\cos^2 \epsilon - \cos^2 \phi}}{\cos \epsilon + \sqrt{\cos^2 \epsilon - \cos^2 \phi}} \right\}}\ ; (1)$$

or

$$E = \frac{H^2 \gamma}{2}(B)\sqrt{(C) + (D)A^2 + (E)A} \ . \quad (1')$$

$$\tan \delta = \frac{\sin \alpha \cos \epsilon + \sin \epsilon \cos(\epsilon - \alpha)A}{\cos \epsilon \cos(\epsilon - \alpha)A}\ ; \ . \quad (1a)$$

or

$$\tan \delta = \frac{\sin \alpha}{\cos(\epsilon - \alpha)A} + \tan \epsilon, \quad \ldots \quad (1'a)$$

where

$$A = \cos\epsilon \, \frac{\cos\epsilon - \sqrt{\cos^2\epsilon - \cos^2\phi}}{\cos\epsilon + \sqrt{\cos^2\epsilon - \cos^2\phi}} \quad . \quad . \quad (d)$$

CASE II. *Surface of earth inclined and* $\alpha = 0$.

$$E = P = \frac{H^2\gamma}{2}\left\{\cos\epsilon \, \frac{\cos\epsilon - \sqrt{\cos^2\epsilon - \cos^2\phi}}{\cos\epsilon + \sqrt{\cos^2\epsilon - \cos^2\phi}} = A\right\}. \quad (2)$$

From Diagram I the values of $A$ can be found for all values of $\phi$ from 0° to 90° and of $\epsilon$ from 0° to 90°, varying by 5°.

$$\delta = \epsilon; \quad . \quad . \quad . \quad . \quad . \quad . \quad (2a)$$

or for all vertical walls the *direction of the earth-pressure is parallel to the surface of the earth.*

CASE III. *The surface of the earth parallel to the surface of repose.*

$$\epsilon = \phi.$$

$$E = \frac{H^2\gamma}{2}\frac{\cos(\phi - \alpha)}{\cos^2\alpha \cos\phi}\sqrt{\sin^2\alpha + \cos^2(\phi - \alpha) + 2\sin\alpha \sin\phi \cos(\phi - \alpha)}. \quad (3)$$

$$\tan\delta = \frac{\sin\alpha + \sin\phi \cos(\phi - \alpha)}{\cos\phi \cos(\phi - \alpha)}. \quad . \quad (3a)$$

CASE IV. *The surface of the earth parallel to the surface of repose and the back of the wall vertical.*

$$\epsilon = \phi \quad \text{and} \quad \alpha = 0.$$

$$E = \frac{H^2\gamma}{2}\cos\phi. \quad . \quad . \quad . \quad . \quad . \quad (4)$$

$$\delta = \phi. \quad . \quad . \quad . \quad . \quad . \quad . \quad (4a)$$

## FORMULAS FOR EARTH-PRESSURE.

CASE V. *The surface of the earth horizontal.*

$$\epsilon = 0.$$

$$E = \frac{H^2 \gamma}{2} \sqrt{\tan^2 \alpha + \tan^4\left(45° - \frac{\phi}{2}\right)}. \quad . \quad (5)$$

$$\tan \delta = \frac{\tan \alpha}{\tan^2\left(45° - \frac{\phi}{2}\right)} . \quad . \quad . \quad . \quad . \quad . \quad . \quad (5a)$$

CASE VI. *The surface of the earth horizontal and the back of the wall vertical.*

$$\epsilon = 0 \quad \text{and} \quad \alpha = 0.$$

$$E = \frac{H^2 \gamma}{2} \tan^2\left(45° - \frac{\phi}{2}\right) \quad . \quad . \quad . \quad (6)$$

$$\delta = 0. \quad . \quad . \quad . \quad . \quad . \quad . \quad . \quad . \quad . \quad (6a)$$

CASE VII. *Fluid pressure.*

$$\epsilon = \phi = 0.$$

$$E = \frac{H^2 \gamma}{2 \cos \alpha}. \quad . \quad . \quad . \quad . \quad . \quad . \quad (7)$$

$$\delta = \alpha. \quad . \quad . \quad . \quad . \quad . \quad . \quad . \quad . \quad (7a)$$

### GRAPHICAL CONSTRUCTIONS FOR DETERMINING THE THRUST OF EARTH.

The following constructions are perfectly general, and apply to *any plane* within a mass of earth. When applied

26 FORMULAS FOR EARTH-PRESSURE.

for determining the thrust of earth against a *retaining-wall*, $\alpha$ and $\epsilon$ are taken as *positive*.

### * Construction (a).

Let $BE$ represent the surface of the earth and $BA$ the back of the wall. Draw $AF$ parallel to $BE$, and at any point $D$ in $AF$ lay off $DF$ equal to the vertical $DE$. Draw

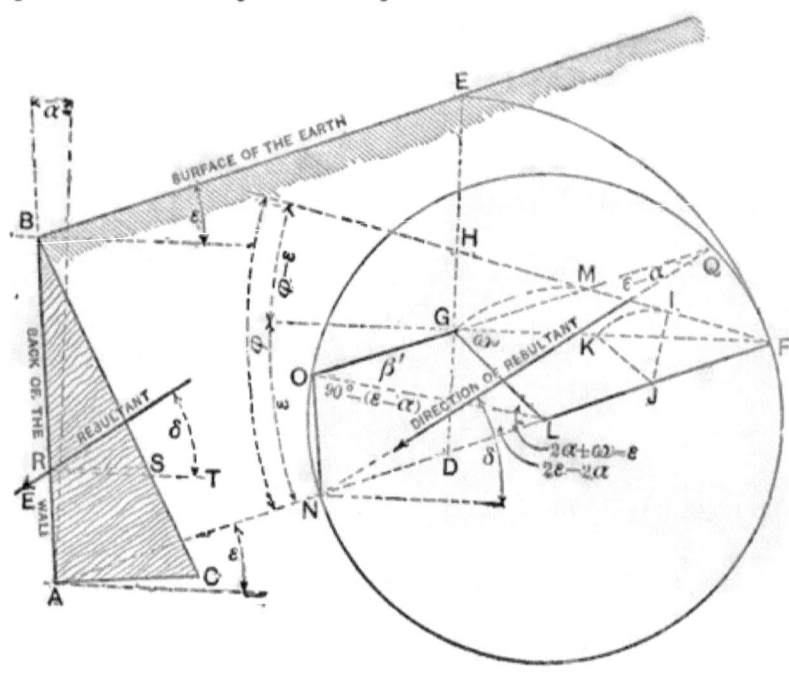

FIG. 10.

$FG$ horizontal, and $FH$, making the angle $\phi$ with $DF$. With any point $J$ in $DF$ describe the arc $KI$ tangent to $HF$ at $I$ cutting $FG$ at $K$, and draw $GL$ parallel to $KJ$; with $L$ as a centre and $LF$ as radius, describe the circumference $FQON$ cutting $AD$ at $N$. Through $N$ draw $NO$

---

* See "Theorie des Erddruckes auf Grund der neueren Anschauungen," by Prof. Weyrauch, 1881.

# FORMULAS FOR EARTH-PRESSURE.

parallel to $AB$ cutting the circumference $FQON$ at $O$; at $A$ draw $AC$ equal to $OG$ and normal to $AB$; the area of the triangle $ABC$ multiplied by $\gamma$ will be the thrust of the earth on the wall.

To determine the direction of the thrust $E$, prolong $OG$ to $Q$; then $QN$ will be the direction of the thrust.

This thrust acts on the wall at $\frac{2}{3}AB$ below $B$.

### * Construction (b).

Let $BQ$ represent the surface of the earth, and $BA$ the back of the wall. Draw $AD$ parallel to $BQ$, and at any

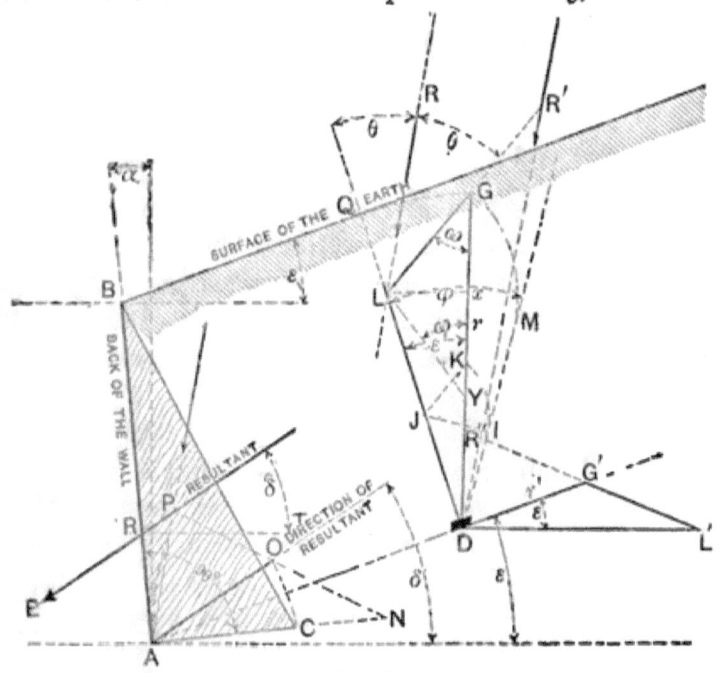

Fig. 11.

point $D$ in $AD$ draw the vertical $DG$ equal to the normal $DQ$; draw $DM$ making the angle $\phi$ with the normal $DQ$.

---

* This construction follows directly from Rankine's Ellipse of Stress. See Rankine's Applied Mechanics.

At any point $J$ in $DQ$ as a centre, describe the arc $IK$ tangent to $DM$ cutting $DG$ at $K$, and draw $GL$ parallel to $JK$. Bisect the angle $QLG$, and at $A$ draw $AP$ parallel to $LR$. At $A$ draw $AN$ normal to $AB$ and equal to $DL$; with $N$ as a centre and $AN$ as radius, describe an arc $AP$ cutting $AP$ at $P$; connect $P$ and $N$, and make $NO$ equal to $LG$; with $A$ as a centre and $AO$ as a radius, describe the arc $OC$ cutting $AN$ at $C$; then the area of the triangle $ABC$ multiplied by $\gamma$ will be the thrust against the wall. The direction of this thrust is parallel to $AO$ and it is applied at $\frac{2}{3}AB$ below $B$.

The constructions (*a*) and (*b*) give identical results in every case.

## Stability of Trapezoidal Walls

As the majority of walls retaining earth are trapezoidal in section, the stability of such walls alone will be considered. If other forms occur in practice they can be divided into trapezoidal sections with horizontal beds, and the stability of each considered, commencing with the upper section.

Walls having the rear faces in the form of steps can usually be considered as trapezoidal in section by replacing the stepped portion by a straight line which approximately bisects each step. If the front faces are stepped they can be treated in a similar manner.

In case the front face of the wall is curved in profile, the curve may be replaced by straight lines which are chords of the curve, thus binding the section into as many trapezoids as there are chords.

It will be assumed that the direction and magnitude of the earth-pressure is known, that the position and extent of the back of the wall, and the width of the top are given, to determine the width of the base for stability against overturning, sliding, and crushing of the material.

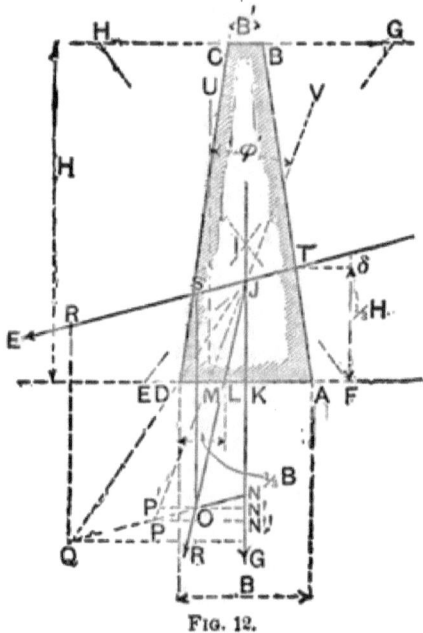

Fig. 12.

*Stability against Overturning.*—Let $ABCD$, Fig. 12, represent a section of a trapezoidal wall, $TR$ the direction of the earth-thrust, $JG$ the vertical passing through the centre of gravity of the wall, and $JO$ the direction of the resultant pressure on the base $AD$ caused by $E$ and $G$.

As long as $R$ cuts the base $AD$, the wall will be stable against overturning. When $R$ takes the direction $JQ$, the wall may be said to be on the point of overturning; then the factor of safety against overturning is $\dfrac{QN}{ON}$, where $ON$ is the actual value of $E$, and $QN$ the value of $E$ required to make the resultant $R$ pass through $D$.

*Stability against Sliding.*—Since the wall will not slide

along the surface $DA$ until the resultant $R$ makes an angle with the normal to $DA$ greater than the angle of friction $\phi'$, the factor of safety against sliding can be obtained as follows: Draw $JP$ making the angle $JMU = \phi'$; then the factor of safety against sliding is $\dfrac{PN}{ON}$, where $PN$ is the force required in the direction of $E$ to make $R$ make the angle $\phi'$ with the normal to $AD$, and $ON$ the actual value of $E$.

*Stability against the Crushing of the Material.*—In ordinary practice walls for retaining earth are not of sufficient height to cause very large pressures at their bases, but it is necessary to consider the subject on account of the tendency of the bed-joints to open under certain conditions.

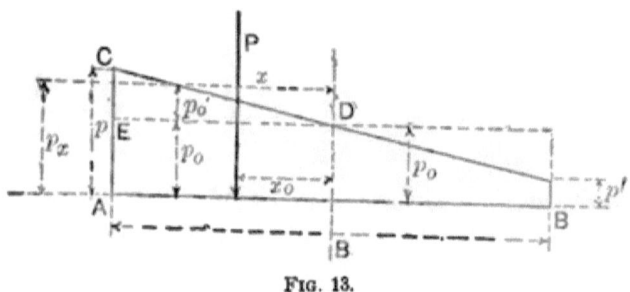

Fig. 13.

Let $AB$, Fig. 13, represent any bed-joint in the wall, $P$ the vertical resultant pressure upon the joint, and $x_0$ the distance of the point of application from the centre of the joint.

The intensity of $P$ at any point can be considered as composed of a uniform intensity $p_0 = \dfrac{P}{B}$, and a uniformly varying intensity $p_0'$, so that $p_x = p_0 + p_0'$. Let $a$ equal the tangent of the angle $CDE$, then $p_0' = ax$ and $p_x = p_0 + ax$.

The pressure upon a surface ($dx$)—the joint being considered unity in the dimension normal to the page—is

$$p_x dx = p_0 dx + axdx,$$

and the moment of this about $DB$ is

$$(p_0 dx + axdx)x.$$

The algebraic sum of these moments for values of $x$ between the limits $\pm \frac{B}{2}$ must equal $Px_0$, or

$$Px_0 = \int_{-\frac{1}{2}B}^{+\frac{1}{2}B} (p_0 x dx + ax^2 dx).$$

Integrating,

$$a = \frac{12 x_0 P}{B^3} = \frac{12 x_0 p_0}{B^2},$$

and

$$p_x = \frac{B^2 + 12 x x_0}{B^2} p_0,$$

or making $x = \frac{1}{2}B$,

$$p = \left\{ 1 + \frac{6 x_0}{B} \right\} \frac{P}{B};$$

and if $x_0$ be replaced by $\frac{1}{2}B - Q$, where $Q$ is the distance from $A$ to the point where $P$ cuts the base, (Fig. 13,)

$$p = 2 \left( 2 - \frac{3Q}{B} \right) \frac{P}{B},$$

and

$$p' = 2 \left( -1 + \frac{3Q}{B} \right) \frac{P}{B}.$$

If $Q = \frac{1}{3}B$,

$$p' = 0 \quad \text{and} \quad p = 2p_0.$$

from which it is seen that when $R$ cuts the base outside the middle third, **the joint** will have a tendency to open at points which are at a **maximum** distance from $R$ where it cuts the base.

**Therefore in no case** should the resultant pressure be **permitted to** cut the base outside the middle third. This makes it unnecessary to consider the stability against over turning.

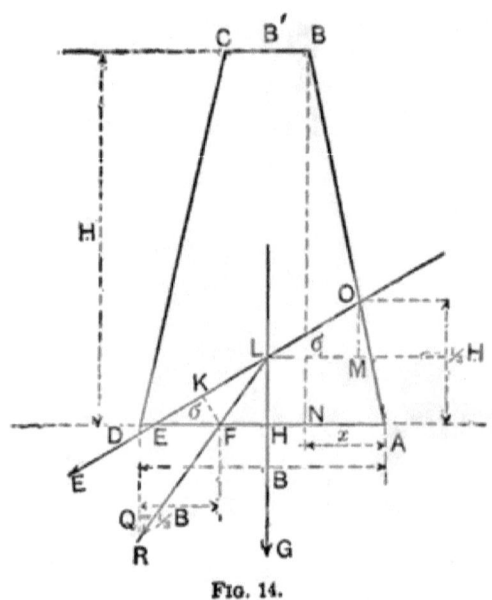

Fig. 14.

Then in designing **a** wall the following conditions must exist for stability:

I. *The resultant $R$ must cut the base for* **stability** *against overturning.*

II. *The resultant $R$ must not make an angle with the normal to the base of the wall greater than the angle of friction $\phi'$.*

III. *The resultant R must not cut the base outside of the middle third, in order that there may be no tendency for the bed-joints to open.*

The above three conditions apply to any bed-joint of the wall; but if they are satisfied at the base and the wall has the section shown in Fig. 14, it will not be necessary to consider any joints above the base unless the character of the stone or the bonding is different.

*Determination of the width of the base of a retaining-wall under the condition that R cuts the base at a point $\frac{1}{3}B$ from the toe of the wall.*

Let $H$, $B'$, $x$, $\delta$, and $E$ be given to determine $B$.

From Fig. 14,

$$KF = \frac{x}{3}\sin\delta + \frac{H}{3}\cos\delta - \frac{2B}{3}\sin\delta,$$

$$HD = \frac{2B^2 + 2BB' - Bx - 2B'x - B'^2}{3(B+B')},$$

$$HF = HD - \frac{B}{3} = \frac{B^2 + BB' - Bx - 2B'x - B'^2}{3(B+B')}.$$

For equilibrium

$$E(KF) = G(HF) = \frac{B+B'}{2} HW(HF).$$

Substituting the values of $KF$ and $HF$ in the above and reducing, it becomes

$$B^2 + B\left(\frac{4E}{HW}\sin\delta + B' - x\right)$$
$$= \frac{2E}{HW}(H\cos\delta + x\sin\delta) + 2B'x + B'^2, \quad . \quad (8)$$

which is the general equation for the width of the base of a trapezoidal wall.

For a rectangular wall $B' = B$.
For a triangular wall $B' = 0$.
For a wall with a vertical front $B' + x = B$ or $B' = B - x$.
For a wall with a vertical back $x = 0$.

Equation (8) is easily transformed to satisfy the requirements of special cases.

The width of the base can be found graphically by assuming a value for $B$ and finding the value of $Q$; if it is less than $\frac{1}{3}B$ another value of $B$ must be assumed, and so on until $Q$ is equal to or greater than $\frac{1}{3}B$.

## Formulas for Trapezoidal and Triangular Walls.

Formulas for the width of the base of trapezoidal walls under the condition that the resultant $R$ cuts the base at a point distant from the toe of the wall equal to one third the width of the base, or $Q = \frac{1}{3}B$.

Case I. *The general case in which the back of the wall is inclined, and $E$ makes an angle with the horizontal.*

$$B^2 + B \left(\frac{4E}{HW} \sin \delta + B' - x\right)$$
$$= \frac{2E}{HW}\left(H \cos \delta + x \sin \delta\right) + 2B'x + B'^2 \quad (8)$$

Case II. *The back of the wall vertical.*

$$x = 0.$$

$$B^2 + B\left(\frac{4E}{HW} \sin \delta + B'\right) = \frac{2E}{W} \cos \delta + B'^2. \quad (9)$$

FORMULAS FOR EARTH-PRESSURE.   35

CASE III. *The back of the wall vertical and the thrust normal to the wall.*

$$x = 0 \quad \text{and} \quad \delta = 0.$$

$$B^2 + BB' = \frac{2E}{W} + B'^2. \quad \ldots \quad (10)$$

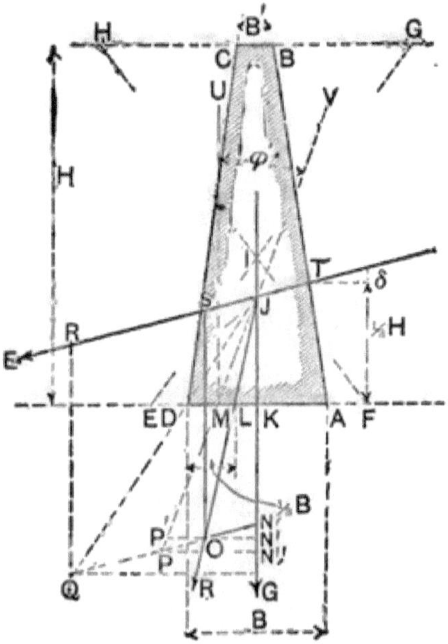

Fig. 15.

If $B = B'$ and $x = 0$, the section of the wall is a rectangle, and (9) becomes

$$B^2 + B\frac{4E}{HW}\sin\delta = \frac{2E}{W}\cos\delta, \quad \ldots \quad (9a)$$

and (10) becomes

$$B = \sqrt{\frac{2E}{W}}. \quad \ldots \quad \ldots \quad (10a)$$

**Formulas for the width of the base of** triangular walls under the condition that the resultant $R$ cuts the base at a point distant from the toe of the wall equal to one third the width of the base, or $Q = \tfrac{1}{3}B$.

CASE I. *The general case in which the back of the wall is inclined, and $E$ makes an angle with the horizontal.*

$$B^2 + B\left(\frac{4E}{HW}\sin\delta - x\right) = \frac{2E}{HW}(H\cos\delta + x\sin\delta). \quad (11)$$

CASE II. *The back of the wall vertical.*

$$\alpha = 0.$$

$$B^2 + B\left(\frac{4E}{HW}\sin\delta\right) = \frac{2E}{W}\cos\delta. \quad . \quad . \quad (12)$$

CASE III. *The back of the wall vertical, and the thrust normal to the wall.*

$$x = 0 \quad \text{and} \quad \delta = 0.$$

$$B = \sqrt{\frac{2E}{W}}. \quad . \quad . \quad . \quad . \quad . \quad (13)$$

The above formulas do not contain the condition that $R$ shall not make an angle greater than $\phi'$ with the normal to the base of the wall.

From Fig. 15,

$$\tan\phi' \gtreqless \frac{E\cos\delta}{G + E\sin\delta} = \tan LJK, \quad . \quad . \quad (14)$$

which expresses the condition under which the wall will not slide.

# FOUNDATIONS FOR WALLS RETAINING EARTH.

The design of the foundations for retaining-walls has received but little attention by writers upon engineering subjects, and the practical engineer has not published to any great extent examples of the foundations he has employed under the countless number of walls erected along railways, highways, canals, etc.

As the designing of foundations resting upon earth, for walls retaining earth, introduces several features which do not influence the ordinary cases of foundations, it will be best to make a special investigation for such conditions.

The intensity of the foundation pressure upon the earth is seldom uniform, due principally to the pressure of the earth behind the wall and foundation tending to overturn the structure as a whole; this being the case, evidently the maximum intensity upon the earth must not be large enough to heave the earth, and the minimum intensity must not be so small that the earth may heave the foundation.

If the foundation be so designed that neither it nor the earth can be heaved, the structure may yet fail by sliding forward. This can only be resisted by the abutting power of the earth in front of the foundation and the friction upon the base of the foundation. Usually, however, if there is no danger of any movement in a vertical plane, there is little or no danger of any movement in a horizontal direction.

As in any structure good judgment must enter into the design, the formulas which will be demonstrated must be used as guides only. These formulas will depend upon the angle of repose $\phi$ of a homogeneous granular mass, and the specific gravity of this mass. For ordinary earths for which the weights and angles of repose are known the results obtained by the use of the formulas will compare very favorably with those obtained from examples of the best practice.

*Depth of Foundations.*—Given the angle of repose $\phi$ of any earth, to determine the depth to which it is necessary to sink a foundation to support a given load. The surface of the earth is assumed to be horizontal.

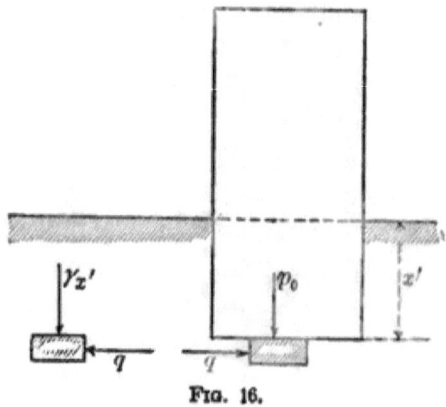

FIG. 16.

CASE I. *When the intensity of the pressure on the base of the foundation is uniform.*

In Fig. 16, let $p_0$ represent the intensity of the pressure on the base of the foundation.

Now when the masonry is about to sink (see Eq. (c)),

$$\frac{p_0}{q} = \frac{1 + \sin \phi}{1 - \sin \phi} \quad \text{or} \quad q = p_0 \cdot \frac{1 - \sin \phi}{1 + \sin \phi}.$$

If $x'$ represents the depth to which the foundation extends below the surface of the earth and $\gamma$ the weight of a cubic foot of earth, then $\gamma x'$ equals the vertical intensity of the earth-pressure on a plane at the depth of the lowest point of the foundation.

When the wall is on the point of sinking, the earth must be on the point of rising, or

$$\frac{q}{\gamma x'} = \frac{1 + \sin \phi}{1 - \sin \phi},$$

or

$$p_0 = \gamma x' \left\{ \frac{1 + \sin \phi}{1 - \sin \phi} \right\}^2. \quad \ldots \quad (15)$$

In any case $p_0$ must not have a greater value than that obtained from (15)—

$$x' = \frac{p_0}{\gamma} \left\{ \frac{1 - \sin \phi}{1 + \sin \phi} \right\}^2 = \frac{p_0}{\gamma} \tan^4 \left( 45° - \frac{\phi}{2} \right). \quad (16)$$

The value of $x'$ as obtained from (16) is the least allowable value consistent with equilibrium. Since $x'$ is a function of $\tan^4 \left( 45° - \frac{\phi}{2} \right)$, care must be taken that $\phi$ is assumed at its least value. As $\phi$ becomes smaller the value of $x'$ increases rapidly.

CASE II. *When the intensity of the pressure on the base is uniformly varying.*

Let $p$ represent the maximum intensity of the pressure on the earth and $p'$ the minimum intensity; then for equilibrium $p$ must not exceed the value obtained from the following equation (see 15):

$$p = x' \gamma \left\{ \frac{1 + \sin \phi}{1 - \sin \phi} \right\}^2. \quad \ldots \quad (17)$$

For any assumed depth $x'$ the maximum value of $p$ can be

found from (17). For any assumed breadth $B''$ of the foundation the value of $p$ due to the resultant pressure upon the base of the foundation can be found from the formulas on page 31, when the value of $x_0$ has been determined; this value must not be greater than the value of $p$ found from (17), or the masonry will heave the earth.

In order that the earth may not heave the masonry, $p'$ must not be *less* than the value obtained from the following formula:

$$p' = x'\gamma \left\{ \frac{1 - \sin \phi}{1 + \sin \phi} \right\}^2 \quad \ldots \quad (18)$$

Then

$$p_0 = \frac{p + p'}{2} = \frac{x'\gamma}{2} \left\{ \left(\frac{1 + \sin \phi}{1 - \sin \phi}\right)^2 + \left(\frac{1 - \sin \phi}{1 + \sin \phi}\right)^2 \right\}, \quad (19)$$

which expresses the *maximum* value $p_0$ can have for the equilibrium of the earth and the masonry.

In order that $p'$ may never be less than the value obtained from (18), the resultant pressure upon the base of the foundation must cut the base within a certain distance of its centre. If $x_0$ be this distance, then (page 31)

$$p' = x'\gamma \left\{ \frac{1 - \sin \phi}{1 + \sin \phi} \right\}^2 = \left\{ 1 - \frac{6x_0}{B''} \right\} p_0. \quad (20)$$

Substituting the value of $p_0$ from (19) and solving for $x_0$,

$$x_0 = \frac{B''}{6} \left\{ \frac{X - Y}{X + Y} \right\}, \quad \ldots \quad (21)$$

where

$$*X = \left(\frac{1 + \sin \phi}{1 - \sin \phi}\right)^2 \quad \text{and} \quad Y = \left(\frac{1 - \sin \phi}{1 + \sin \phi}\right)^2.$$

---

\* Tabulated values of $X$ and $Y$ are given on page 72.

FOUNDATIONS FOR WALLS RETAINING EARTH.  41

*Depth of foundations when the surface of the earth has different elevations on opposite sides of the structure.*

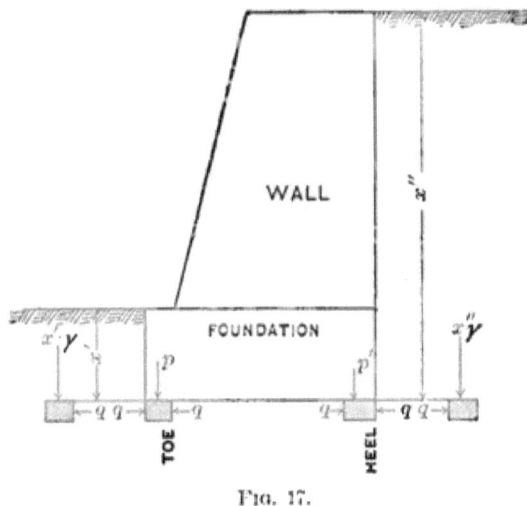

FIG. 17.

This case is illustrated in Fig. 17.  From (17) and (18) for equilibrium

$$p \leqq x'\gamma \left\{ \frac{1+\sin\phi}{1-\sin\phi} \right\}^2 \quad \ldots \quad (22)$$

and

$$p' \geqq x''\gamma \left\{ \frac{1-\sin\phi}{1+\sin\phi} \right\}^2 \quad \ldots \quad (23)$$

Combining (22) and (23) in the value of $p_0$,

$$p_0 = \frac{p+p'}{2} = \frac{\gamma}{2} \left\{ x'\left(\frac{1+\sin\phi}{1-\sin\phi}\right)^2 + x''\left(\frac{1-\sin\phi}{1+\sin\phi}\right)^2 \right\}. \quad (24)$$

Having assumed the values of $\gamma$ and $\phi$ for any particular case, the above formulas determine the permissible magnitudes of the intensities at the heel and toe of the foundation for any depth. The breadth of the base of the foundation may now be assumed, and the actual intensities compared with those permissible; if $p$ is too large or $p'$ too small, another trial must be made. Usually one or two trials are sufficient. If one prefers to compute the width of the base of a trapezoidal foundation, the formula given below can be employed.

*Determination of the breadth $B''$ of a trapezoidal foundation for a given loading and a maximum intensity $p$ at the toe.* (Back of foundation vertical.)

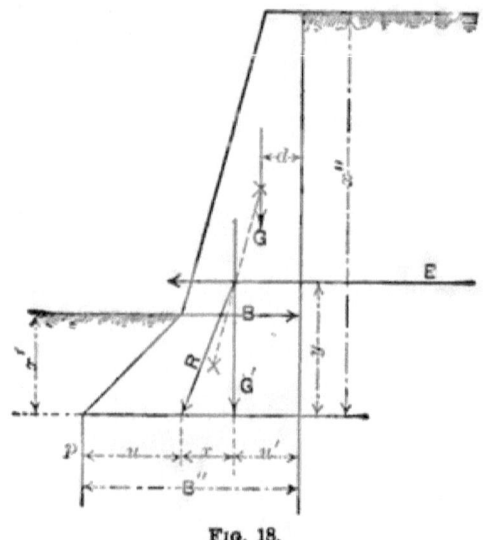

Fig. 18.

Let $G$ = total vertical weight supported by top of foundation;
$E$ = thrust of earth;
$p$ = maximum intensity of pressure at toe of foundation as found from (22);
and $B''$ = breadth of base of foundation.

## FOUNDATIONS FOR WALLS RETAINING EARTH. 43

Then

$$B'' = \frac{-G + \sqrt{p(6dG + x'B^2W + 6Ey) + G^2}}{p}. \quad (25)$$

The foundation can nearly always be designed as a trapezoid having a vertical back, and then if necessary the batter in front can be stepped. For walls under twenty feet in height, retaining material which will assume a slope of $1\frac{1}{2}$ to 1, the most economical foundation is rectangular in section if the base must be four feet deep to escape the action of frost. Where frost need not be considered, of course more shallow and broader foundations can be employed.

*Abutting Power of Earth.*—Let the surface of the earth be horizontal and the body pushing the earth have a vertical face; then at the depth $x'$ the maximum horizontal pressure per unit of area is (see Case I above)

$$q = x'\gamma \frac{1 + \sin \phi}{1 - \sin \phi},$$

and since $q$ varies directly as $x'$, the total thrust $P$ which the earth is capable of resisting is

$$P = \frac{(x')^2 \gamma}{2} \frac{1 + \sin \phi}{1 - \sin \phi}. \quad . \quad . \quad (26)$$

*Bearing Power of Earth.*—The bearing power or the intensity of the pressure which earth can resist depends not only upon the character of the earth, but upon the depth to which the foundation is extended, as shown by the formulas for $p$ given above. For example, the foundation may be very broad and shallow or quite narrow and deep. The intensity of the pressure in the first case being considerably smaller than in the second, and both conditions fulfilling the conditions of stability. It appears then that the bearing powers of earth given by various writers must be employed with caution, unless the conditions upon which the values were based are known.

## APPLICATIONS.

The determination of the earth-pressure by the preceding formulas and graphical constructions is a very simple operation when the angle $\phi$ has been determined or assumed. That care and judgment be used in assuming the value of $\phi$ is very important, since a change of a few degrees in the value of $\phi$ sometimes causes a large change in the value of $E$. An inspection of Diagram I shows that the value of the coefficient $A$ increases very rapidly as $\phi$ decreases.

When the earth to be retained contains springs, the bank must be thoroughly drained if it is to be retained by an economical tight wall; if it is not drained, the angle $\phi$ will be likely to become very small as the earth becomes wet.

When the location of the earth to be retained is subjected to jars, the value of $\phi$ will be decreased.

Hence, in assuming the value of $\phi$, the engineer must be sure that the value assumed will be the least value which, in his judgment, it is likely to have.

In constructing the wall the judgment and authority of the engineer must again be exercised in order that the wall be constructed as designed.

In all cases, to insure perfect drainage between the back

## FOUNDATIONS FOR WALLS RETAINING EARTH. 45

of the wall and the earth, numerous "weep-holes" should be provided in the body of the wall, or proper arrangements made to carry away the water at the base of the wall. To facilitate drainage, the backing resting against the wall should be sand or gravel.

In no case should water be permitted to get under the foundation of the wall, neither should the earth in front of the wall be allowed to become wet.

In cold localities the back of the wall near the top should have a large batter to prevent the frost from moving the top courses of stone. As a guard against sliding, the courses of the wall should have very rough beds. The strength of a wall is increased the nearer it approaches a monolith.

Care should be taken to have the foundation broad and deep enough to prevent sliding and upheaving of the earth in front. In clay the foundation should be deep, while in sand or gravel it may be broad and shallow.

The following examples illustrate the application of the formulas:

Ex. 1. Design a trapezoidal wall of sandstone, weighing 150 lbs. per cubic foot, having a width of 3 ft. on top, a height of 30 ft., and the back inclining forward 5°, to retain a bank of sand sloping upward at an angle of 20°.

### Data.

$\gamma = 100$ lbs., $W = 150$ lbs.; $\epsilon = 20°$, $\phi = 39°$, $\alpha = 5°$; $H = 30$ ft., $B' = 3$ ft., $x = 2.63$ ft.

1°. *Graphical determination of the values of $E$ and $\delta$.*

The graphical solution of the problem is shown in Fig. 19, where $E$ is found to equal 15,000 pounds. $\delta$ lies between 35° and 36°.

2°. *Algebraic determination of E and δ.*

$$E = \frac{H^2\gamma}{2}(B)\sqrt{(C)+(D)A^2+(E)A}. \quad \ldots \quad (1')$$

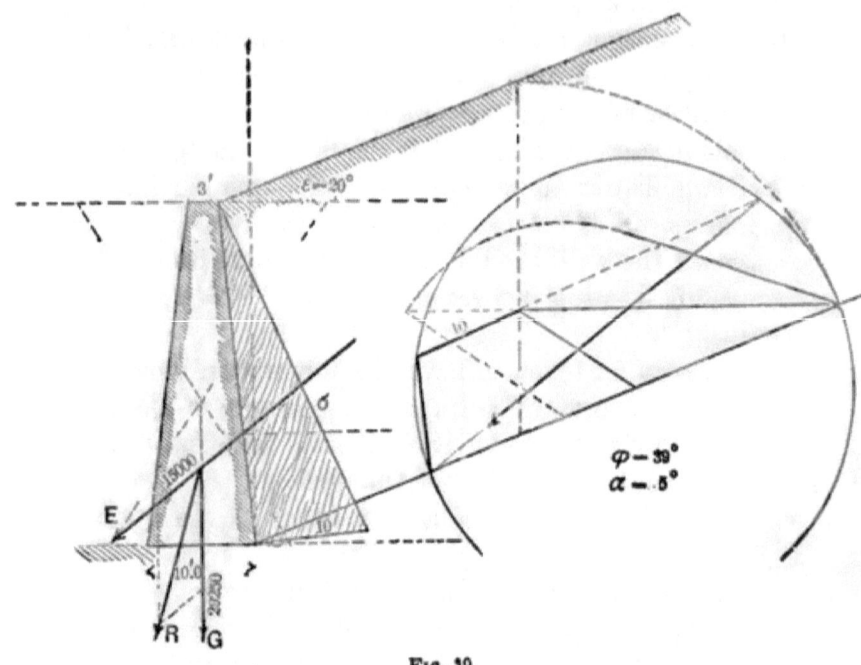

Fig. 12.

Substituting the values of *B, C, D,* and *E* as given in the tables, and that of *A* as given by Diagram I, this becomes

$$E = \frac{900 \times 100}{2}(1.036) \times$$

$$\sqrt{(0.008)+(1.057)(0.264)^2+(0.061)0.264},$$

$$E = 45{,}000\,(1.036)\sqrt{0.098} = 14{,}500 \text{ lbs.}$$

$$\tan \delta = \frac{\sin \alpha}{\cos(\epsilon - \alpha)A} + \tan \epsilon, \quad \ldots \quad (1'a)$$

FOUNDATIONS FOR WALLS RETAINING EARTH. 47

$$\tan \delta = \frac{0.087}{0.966(0.264)} + 0.364,$$

$$\tan \delta = 0.705 = \tan 35° \; 11', \text{ about.}$$

3°. *Algebraic determination of the value of $B$ under the assumption that $Q = \frac{1}{4}B$.*

$$B^2 + B \left\{ \frac{4E}{HW} \sin \delta + B' - x \right\}$$
$$= \frac{2E}{HW} \left\{ H \cos \delta + x \sin \delta \right\} + 2B'x + B'^2. \quad . \quad (8)$$

$$B^2 + B \left\{ \frac{4 \times 14500}{30 \times 150} 0.576 + 3 - 2.63 \right\}$$
$$= \frac{2 \times 14500}{30 \times 150} \{30 \times 0.817 + 2.63 \times 0.576\} + 6 \times 2.63 + 9,$$

$$B^2 + 7.79 B = 172.53,$$

$$B = -3.89 \pm \sqrt{172.53 + 3.9^2};$$

$$\therefore B = 13.69 - 3.89 = 9.80 \text{ ft.};$$

or, practically, 10 feet is the required width of the base.

4°. *To determine if the wall will slide on a foundation of sandstone.*

From (14),

$$\tan \phi' \geqq \frac{E \cos \delta}{G + E \sin \delta}.$$

Taking $B = 10$ ft., $G = \dfrac{10 + 3}{2} \cdot 30 \times 150 = 29250$ lbs.

$\delta = 35° 11'$, $\cos \delta = 0.817$, and $\sin \delta = 0.576$, then

$$\frac{E \cos \delta}{G + E \sin \delta} = \frac{14500 \times 0.817}{29250 + 14500 \times 0.576} = 0.315.$$

From Table II, the value of $\tan \phi'$ for masonry is 0.6 to 0.7; hence there is no danger of the wall sliding on the foundation.

According to the *Engineering News* formula the base of this wall would be $\frac{3}{8}H$ "plus a few inches for good luck," or about 13 feet, and by the old rule of one third the height 10 feet.

**Ex. 2. Design** a trapezoidal **wall** of sandstone weighing **150 lbs. per cubic foot,** having a width **of 3 ft.** on top, a height **of 30 ft., and the back inclining** backward **15°, to retain a bank of sand sloping upward at an angle of 30°.**

*Data.*

$\gamma = 100$ lbs., $W = 150$ lbs.; $\epsilon = 30°$, $\phi = 33°$, $\alpha = -15°$; $H = 30$ ft., $B' = 3$ ft., $x = 8$ ft.

1°. *Graphical determination of the values of $E$ and $\delta$.*

In Fig. 19, let $BG$ represent the surface of the earth, and $AB$ the back of the wall. Draw $AF$ parallel to $BG$, and from any point $D'$ in $AF$ lay off $D'F$ equal to the vertical $D'G$, and draw $FL$ horizontal; lay off the angle $IFD' = \phi = 33°$, and locate the point $M$ in $D'F$ so that if an arc be described with $M$ as a centre and $LM$ as a radius the arc will be tangent to $IF$; then with $M$ as a centre and $MF$ as a radius, describe the circumference $FHJ$ and draw $JH$

# FOUNDATIONS FOR WALLS RETAINING EARTH. 49

parallel to $AB$; at $A$ draw $AL$ perpendicular to $AB$ and equal to $HI$. Then

$$\frac{(AB)(AL)}{2}\gamma = \frac{(30.9)(9.6)}{2}100 = 14800 = E.$$

To determine $\delta$, prolong $HI$ to $K$ and draw $KJ$. Then the angle which this line makes with the horizontal is equal to $\delta$, which is 6° to 7° in this case.

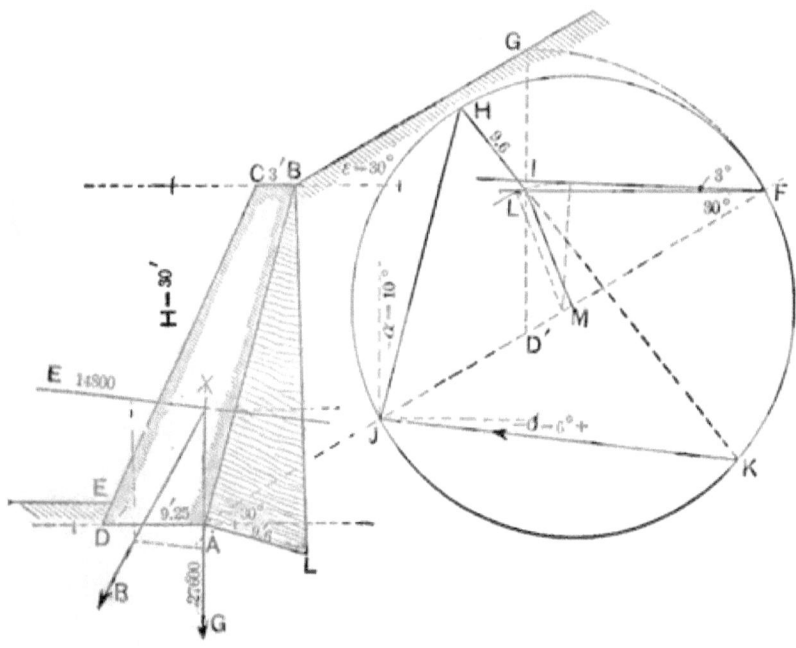

Fig. 20.

**2°.** *Algebraic determination of $E$ and $\delta$.*
Substituting in (1) and remembering that $\alpha$ is negative,

$$E = 45000\,(0.875)\,\sqrt{0.067 + 0.183 - 0.111} = 14600 \text{ lbs.}$$

From (1'a),

$$\tan \delta = \frac{-0.259}{0.707(0.524)} + .577 = -0.123 = \tan(-7°).$$

3°. *Algebraic determination of the value of B under the assumption that* $Q = \tfrac{1}{3}B$.

Substituting the proper values in (11) and remembering that $\alpha$ is negative,

$$B = -4.7 \pm \sqrt{163.44 + (4.7)^2} = 9.0 \text{ ft.}$$

Ex. 3. Determine the dimensions of a brick wall having a vertical back to retain a bank of sand sloping upward at an angle of 20°. $\phi = 30°$, $H = 20'$, $B' = 2'$, $\gamma = 100$.

1°. *Algebraic determination of E and* $\delta$.

Since $\alpha = 0$,

$$E = \frac{H^2 \gamma}{2} A \quad \ldots \ldots \ldots \quad (2)$$

$$E = \frac{400 \times 100}{2} 0.424 = 8480; \text{ say, } 8500 \text{ lbs.}$$

The value of $A$ is readily found from Diagram I.

$$\delta = \epsilon = 20°, \quad \text{since} \quad \alpha = 0.$$

2. *Algebraic determination of the value of B under the condition that* $Q = \tfrac{1}{3}B$.

$$B^2 + B\left\{\frac{4E}{HW}\sin\delta + B'\right\} = \frac{2E}{W}\cos\delta + B'^2. \quad (9)$$

# FOUNDATIONS FOR WALLS RETAINING EARTH. 51

From Table I, $W = 125$ lbs. Then

$$B^2 + B\left\{\frac{4 \times 8500}{20 \times 125}0.342 + 2\right\} = \frac{2 \times 8500}{125}0.940 + 4,$$

or
$$B^2 + 6.65B = 131.84.$$

$$B = -3.36 \pm \sqrt{131.84 + 3.36^2},$$

and
$$B = -3.36 + 11.96 = 7.60 \text{ ft.}$$

Ex. 4. Determine the value of $B$ in Ex. 3 under the assumption that $\epsilon = 0$ (horizontal earth-surface).

$$E = \frac{H^2\gamma}{2}\left\{\tan^2\left(45° - \frac{\phi}{2}\right) = \frac{1 - \sin\phi}{1 + \sin\phi}\right\}, \quad (6)$$

or $E = 20000 (0.333) = 6666$, say 6700 lbs.
Since $\alpha = 0$, and $\epsilon = 0$, $\delta = 0$,

$$B^2 + BB' = \frac{2E}{W} + B'^2; \quad \ldots \quad (10)$$

$$B^2 + 2B = 111.2;$$

$$B = -1 \pm \sqrt{111.2 + 1},$$

and
$$B = -1 + 10.59 = 9.6 \text{ ft.}$$

Ex. 5. Determine the value of $B$ in Ex. 3, under the assumption that $\epsilon = \phi = 30°$.

$$E = \frac{H^2\gamma}{2}\cos\phi = 20000 (0.866) = 17320 \text{ lbs.}$$

From (9),

$$B^2 + B\left\{\frac{4 \times 17320}{20 \times 125}0.5 + 2\right\} = \frac{2 \times 17320}{125}0.866 + 4;$$

52   FOUNDATIONS FOR WALLS RETAINING EARTH.

$$B^2 + 15.86B = 244.05;$$
$$B = -7.93 + \sqrt{244.05 + 7.93^2}.$$

and   $B = -7.93 + 17.52 = 9.6$ ft.

Ex. 6. Determine the resultant pressure against the back of a wall when the surface of the earth carries a load equivalent to 5 feet in depth of sand.

$H = 30$ ft., $\alpha = 10°$, $\phi = 30°$, $\epsilon = 0$, and $\gamma = 100$ lbs.

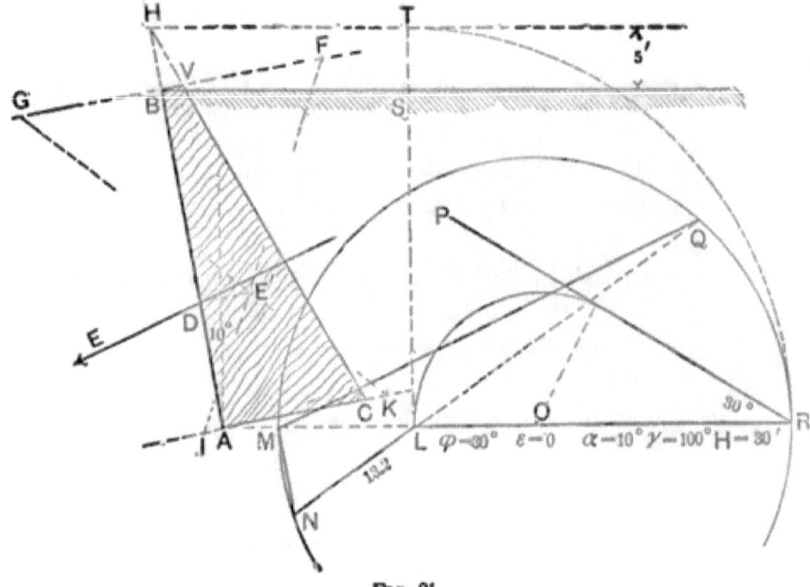

Fig. 21.

*Graphical solution of the problem.*—In Fig. 21, let $BS$ represent the surface of the earth, and $BA$ the back of the wall.

Make $ST = 5$, and draw $HT$ and $BH$. Draw $AR$ parallel to $BS$, parallel to $HT$, and make $LR$ equal to $LT$; lay off the angle $LRP$ equal to 30°; with $O$ as a centre draw an arc passing through $L$ tangent to $PR$, and then with $OR$ as a radius describe the circumference of the circle $RQM$, and at $M$ draw $MN$ parallel to $AH$; at $A$ and normal to $AH$ draw $AC$ equal to $NL$. Then

$$\frac{AC + BV}{2} BA \cdot \gamma = E.$$

The direction of $E$ will be parallel to $QM$.

To determine the point of application of $E$, find the centre of gravity $E'$ of $ABVC$, and draw $E'D$ parallel to $AC$, then $D$ will be the point of application of $E$.

$E'$ can be found as follows: Produce $AC$ and $BV$, make $AI = CK = BV$, $BG = VF = AC$, and join $F$ and $I$ and $G$ and $K$. Then $E'$, the intersection of $FI$ and $GK$, will be the centre of gravity of $ABVC$. $BD$ can be found from the formula

$$BD \cos 10° = \frac{1}{3} \frac{(TL)^3 - 3(TL)(TS)^2 + 2(TS)^3}{(TL)^2 - (TS)^2}.$$

Ex. 7. Determine graphically the value of $E$ when $\epsilon = 0$ and $\alpha = 0$, $\phi$, $\gamma$, and $H$ being given.

In Fig. 22 let $BF$ represent the surface of the earth, and $AB$ the back of the wall. Draw $AL$ parallel to $BF$ and make $IL = IF$; lay off the angle $GLH = \phi$, and at any point $K$ in $LH$ draw $MK$ perpendicular to $HL$, and lay off $MO = MK$; draw $MJ$ parallel to $OI$. Then will the arc $IN$, described with $J$ as a centre and $IJ$ as a radius, pass through $I$ and be tangent to $GL$; with $J$ as a centre and $JL$ as radius describe the circumference $LH$; at $A$ lay off $AC = HI$ and normal to $AB$. Then

$$\frac{AC \times AB}{2} \gamma = E.$$

$E$ is parallel to $BF$ and applied at $D$, $AD$ being equal to $\tfrac{1}{3}AB$.

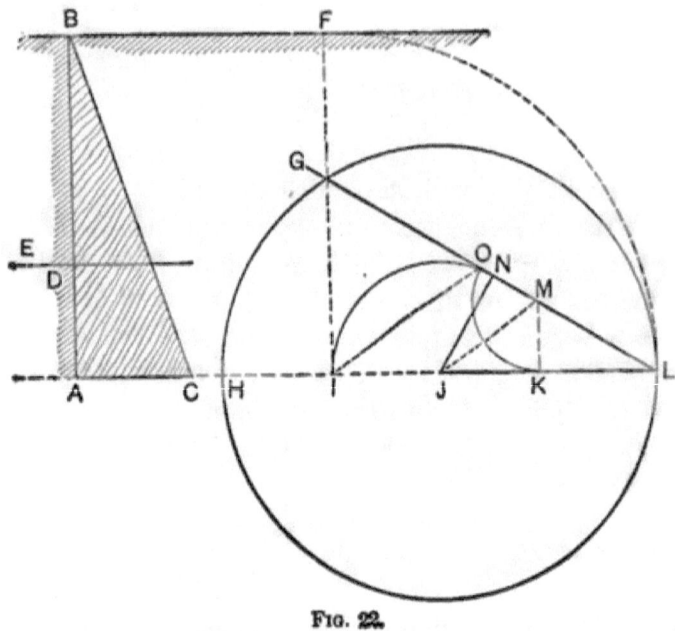

Fig. 22.

Ex. 8. Determine the earth-thrust on the profile shown in Fig. 23, $H$, $\gamma$, $\phi$, and $\epsilon$ being given.

*Graphical solution of the problem.*—Let $BCDEA$ represent the given profile, and let the surface of the earth be horizontal. Prolong $BC$ until it intersects $SA$ in $S$; draw $SR$ normal to $BCS$ and equal to the intensity of the earth-pressure at $S$; connect $B$ and $R$. Then from the middle point of $BC$ draw $GF$ parallel to $SR$; the distance $GF$ multiplied by $\gamma$ will be the average intensity of the earth-pressure on $BC$. In a similar manner the average intensities on $CD$, $DE$, and $EA$ can be found, and hence the total pressures on each determined. The points of application of these resultant pressures, $E_1$, $E_2$, $E_3$, and $E_4$, can be found by the method used in Ex. 6 for finding the centre of gravity of a trapezoid. The directions of

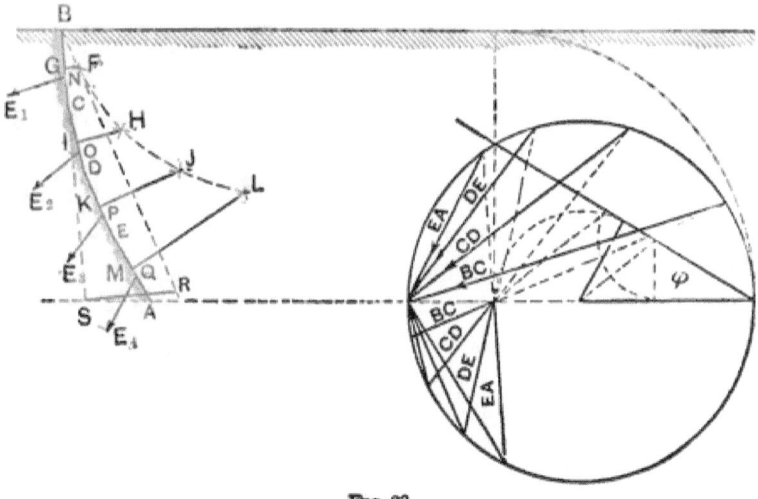

Fig. 23.

$E_1$, $E_2$, $E_3$, and $E_4$ are found from the construction on the right.

Ex. 9. Determine the thrust of the earth against a vertical wall when $\epsilon$ is negative.

For the explanation of this construction, see page 21, Fig. 9.

Ex. 10. From the following data determine $E$, $\delta$, and $Q$:

$\epsilon = 0$, $\phi = 38°$, $\alpha = 10° 23'$; $\gamma = 90$ lbs., $W = 170$ lbs.;

$H = 15$ ft., $B = 6$ ft., $B' = 2$ ft.

Ans. $E = 3037$ lbs., $\delta = 27° 13'$, $Q = 2.2$ ft.

Ex. 11. Determine the dimensions of a trapezoidal wall built of dry, rough granite, having a vertical back and being 20 feet high, to safely retain the side of a sand cut, the surface of the sand being level with the top of the wall. $W = 165$ lbs., $\gamma = 100$ lbs., $\phi = 33°$ 40', $H = 20$ ft., $B' = 2$ ft.

*Ans.* $E = 5734$ lbs., $\delta = 0$, $B = 8$ ft., and $Q = 2.8$ ft., about.

Ex. 12. The same as Ex. 11, with $\alpha = 8°$ instead of $\alpha = 0$.

*Ans.* $E = 6330$ lbs., $B = 8$ ft., and $Q = 2.7$ ft.

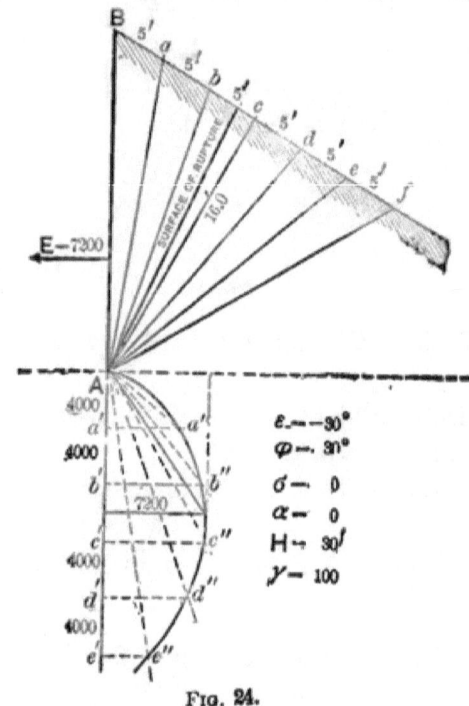

Fig. 24.

Ex. 13. What must be the dimensions of a rubble wall of large blocks of limestone, laid dry, to retain a sand filling which supports two lines of standard-gauge railroad track? (Assume the depth of sand to produce a pressure on the earth equal to that produced by the railroad and trains as 4 feet.)

$H = 15$ ft., $\alpha = 8°$, $\phi = 33° \ 40'$, $\gamma = 100$ lbs., $W = 170$ lbs., $B' = 3.5$ ft.

Ans. $E = 5760$ lbs., $\delta = 18° \ 7'$, $B = 8$ ft., $Q = 2.7$ ft.

Ex. 14. Determine $E$, $\delta$, $B$, and $Q$, when $W = 170$ lbs., $\gamma = 100$ lbs., $\alpha = 8°$, $\epsilon = \phi = 33° \ 40'$, $H = 20$ ft., $B' = 2$ ft.

Ans. $E = 21760$ lbs., $\delta = 32° \ 25'$, $B = 9$ ft., $Q = 3$ ft.

\* Ex. 15. A wall 9 ft. high faces the steepest declivity of earth at a slope of 20° to the horizon; weight of earth 130 lbs. per cubic foot, angle of repose 30°. Determine $E$ when $\alpha = 0$.

Ans. $E = 2187$ lbs.

\* Ex. 16. $\epsilon = 33° \ 42'$, $\phi = 36°$, $H = 3$ ft., $\gamma = 120$ lbs., $\alpha = 0$. Determine $E$.

Ans. $E = 278$ lbs.

\* Ex. 17. $\phi = 25°$, $\epsilon = 0$, $\alpha = 0$, $H = 4$ ft., $\gamma = 120$ lbs., $E = ?$

Ans. $E = 390$ lbs.

\* Ex. 18. $\phi = 38°$, $\epsilon = 0$, $\alpha = 0$, $H = 3$ ft., $\gamma = 94$ lbs., $E = ?$

Ans. $E = 100.5$ lbs.

\* Ex. 19. A ditch 6 feet deep is cut with vertical faces in clay. These are shored up with boards, a strut being put across from board to board 2 feet from bottom, at intervals of 5 feet apart. The coefficient of friction of the moist clay is 0.287, and its weight 120 lbs. per cubic foot. Find the thrust on a strut, also find the greatest thrust which might be put upon the struts before the adjoining earth would heave up.

Ans. $E = 1230$ lbs.
Thrust per strut $= 6128$ lbs.
Greatest thrust $= 19029$ lbs.

---

\* Alexander's Applied Mechanics.

## 58   FOUNDATIONS FOR WALLS RETAINING EARTH.

Ex. 20. Examine the stability of the wall shown in Fig. 25, and design a foundation which will be safe as long as the condition of the earth remains unchanged; the weight of the masonry being 145 pounds per cubic foot, that of earth 100 pounds, and the angle of repose of the earth such that it will stand at a slope of $1\frac{1}{2}$ to 1.

*Stability of the Wall upon the Foundation.*—Replacing the stepped back by the line $BD$, the thrust of the earth is found to be about 9900 pounds. The direction of this force is shown in Fig. 25; since it cuts the base of the wall there is no danger of the structure being overturned, however large $E$ may become.

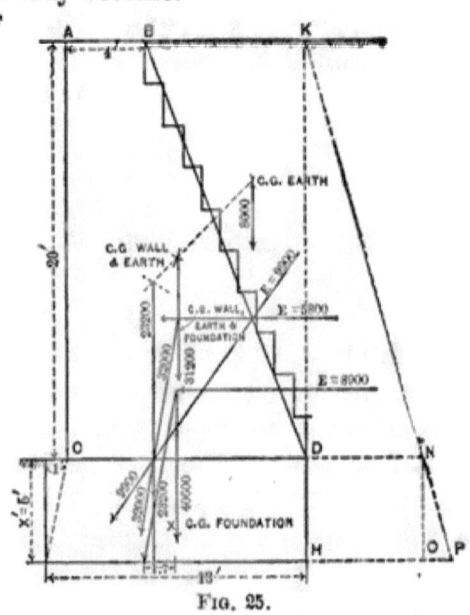

FIG. 25.

Determining the centre of gravity of the wall and also its weight, and combining this with $E$, the resultant pressure upon the base of the wall is found to be about 32,000 pounds. This resultant makes an angle of less than 11

# FOUNDATIONS FOR WALLS RETAINING EARTH. 59

degrees with the normal to the base. Now since for masonry sliding upon masonry the angle of friction is from 31 to 35 degrees (Table II), there is no danger of failure by sliding upon the foundation. Since the resultant cuts the base within the middle third the entire base is subjected to compression, and there will be no tendency for the joints to open at the heel.

Failure by the crushing of the material need not be considered, as the maximum intensity of the pressure upon the base is many times smaller than the ultimate strength of the material. See page 68.

The resultant pressure upon the base can be found also by assuming the earth on the left of the vertical to be supported by the wall, and that the pressure of the earth upon the right of this line acts against the vertical plane $KD$; this pressure is about 5800 pounds, and is horizontal. Combining this force with the weight of the wall and earth on the left of the line $KD$, the resultant pressure upon the base is found to be the *same in magnitude and direction as by the first method*.

*The Foundation.*—The depth of the foundation must be below the action of frost; let this be assumed as 5 feet; then by (22), with $x' = 5$ feet, the *maximum* allowable pressure at the toe of the foundation is about 6000 pounds per square foot, and by (23) the *minimum* allowable pressure is about 200 pounds for $x'' = 25$ feet.

Assuming that the foundation is vertical at the back and trapezoidal in section, the length of the base $B''$ can be found from (25), which will satisfy the condition of maximum pressure at the toe. Letting $p = 5000$ and $x' = 5$, and solving (25), $B''$ is found to be between 12 and 13 feet; say 13 feet.

To determine if this width is sufficient to satisfy all the

conditions of equilibrium, the resultant of all forces acting upon the base must be found.

*The total earth-pressure upon the vertical $HK$ is about 8900 pounds. Combining this with the weight of the wall, earth supported by the wall, and that of the foundation, the resultant vertical pressure is found to be about 40,600 pounds, and is applied within the middle third of the base, about 1.7 feet to the left of the centre.

The intensity of the pressure at the toe is (page 31)

$$p = \left\{ 1 + \frac{6(1.7)}{13} \right\} \frac{40600}{13} = \text{about 5600 pounds,}$$

which is less than the maximum allowable intensity. The intensity at the heel is $p' = 2p_0 - p =$ about 650 pounds, which is greater than the minimum allowable intensity; hence this foundation is sufficient to prevent settlement.

A glance at Fig. 25 is sufficient to show that the foundation will not slide upon the earth even if the movement were not opposed by a force of some 4000 pounds, being the abutting power of the earth in front of the foundation.

The above foundation then fulfils all the conditions of stability, but to allow for contingencies the foundation should be designed under the assumption that $\phi$ may be somewhat smaller than its average value, which is equivalent to broadening the base if the depth remains the same

---

* The pressure against the foundation in front of the wall has been neglected, but can be easily included by taking the area $KHON$ instead of $KHP$.

## PROFILES OF WALLS RETAINING EARTH. 61

## EXAMPLES OF RETAINING-WALL PROFILES.

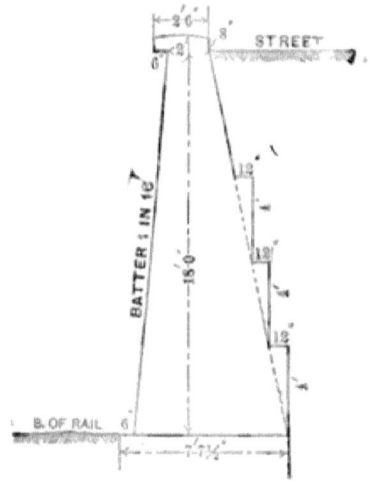

FIG. 26.

A Standard **Profile used for the past twenty years near New York City, where railway tracks have been lowered below the streets.** (*Engineering News*, **1889.**)

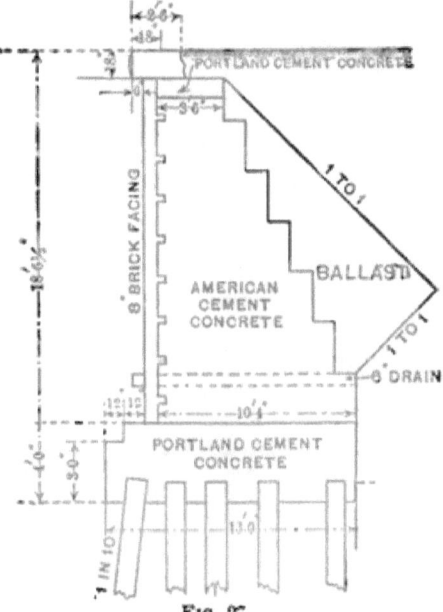

FIG. 27.

Profile of Retaining-wall at Ferdinand Street Bridge, Boston, Mass. (*City Engineer's Report*, 1891.)

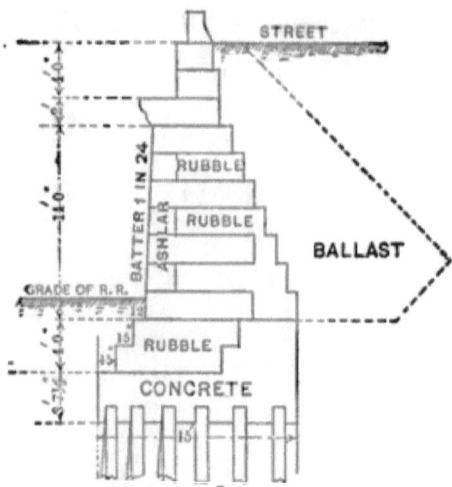

Fig. 28.
Profile of Abutment at Ferdinand Street Bridge, Boston, Mass.
(*City Engineer's Report*, 1891.)

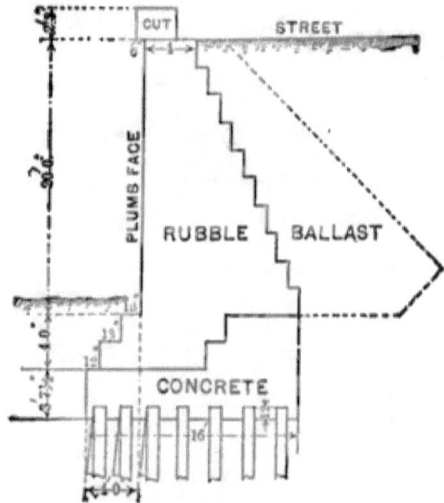

Fig. 29.
Profile of Retaining-wall at Boylston Street Bridge, Boston, Mass.
(*City Engineer's Report*, 1888.)

## PROFILES OF WALLS RETAINING EARTH. 63

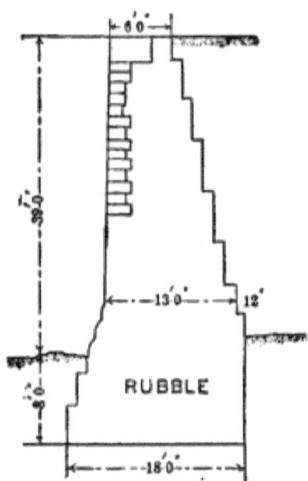

FIG. 30.
Profile of Retaining-wall at Liverpool, England. (*Harcourt*).

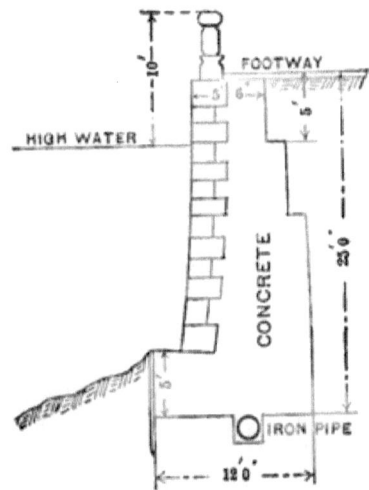

FIG. 31.
Profile of Retaining-wall, Thames Embankment, Chelsea. (*Harcourt.*)

Fig. 32.
Profile of Retaining-wall Thames Embankment, Lambeth. (*Harcourt.*)

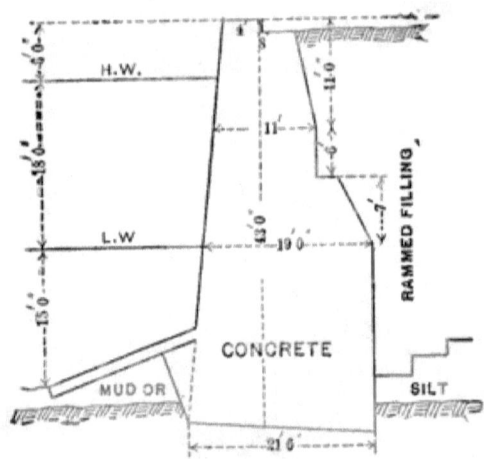

Fig. 33.
Profile of Concrete Retaining-wall at Chatham. (*Harcourt.*)

# PROFILES OF WALLS RETAINING EARTH. 65

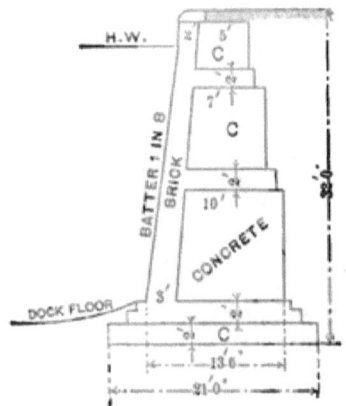

Fig. 34.
Profile of Retaining-wall at Millwall. (*Harcourt.*)

# FOUNDATIONS.

The proper proportions of foundations to suit different conditions have been the results of experience principally, though theory enters into their design in many ways. Under certain logical assumptions, the offsets of wood, iron, or stone foundation courses can be as accurately determined as the stresses in any beam subjected to cross-bending. The strengths of various materials which enter into the construction of foundations have been fairly well determined experimentally, so that the allowable intensities of the pressures, and consequently the areas of the foundation courses, can be accurately determined. There remains the most difficult portion to be decided, namely, the proper intensity of the pressure upon the earth which must support the load. Under certain assumptions this can be computed, but the best of judgment must be exercised in making the assumptions upon which calculations are based.

*Whenever possible, the intensity of the pressure upon the earth should be uniform under all parts of the structure* (assuming the earth to be homogeneous), *and the foundations extend to the same depth.* Theoretically, a greater intensity is allowable at a greater depth, but practically this may lead to unequal settlement, due to the compressibility of the earth, which theory does not take into account.

### Foundations upon Rock.

In preparing a bed for the structure to be erected all loose and decayed parts of the rock must be removed, and the surface made as nearly horizontal as practicable; when the surface is inclined, it may be cut into steps with horizontal and vertical faces; if holes exist, they may be filled with concrete. In some cases a proper surface for supporting the proposed structure can be secured by covering the rock surface with a layer of concrete, which may vary from a few inches to two or more feet in thickness. (Figs. 39 and 42.)

The *maximum* intensity of the pressure upon a rock foundation should not exceed *one sixth* the crushing strength of the rock for a steady and uniform load, or one tenth the crushing strength for a load due to the weight of the structure plus a varying load such as is caused by wind or earth pressure.

In no case should any portion of the horizontal joints be subjected to tension. The maximum deviation of the centre of pressure from the centre of gravity of the base section, when the section is a symmetrical figure, can be found from the formula

$$x_0 = \frac{I}{Ay}, \text{ (Rankine)};$$

where $x_0$ = the maximum deviation sought;

$I$ = the moment of inertia of the section relative to an axis perpendicular to the direction in which the maximum deviation is sought;

and $y$ = the distance from the centre of gravity of the section to the edge furthest from the centre of pressure measured along an axis passing through the centre of pressure and the centre of gravity.

Following are the more common sections of foundations with the corresponding values of $x_0$:

Rectangle...$A = bh$,　　$x_0 = \frac{1}{6}b$;
Circle.....$A = \pi r^2$,　　$x_0 = \frac{1}{8}d$;

Hollow rectangle:

$$A = bh - b'h', \quad x_0 = \frac{b}{6}\left(1 - \frac{b'h'^2}{bh^2}\right) \div \left(1 - \frac{b'h'}{bh}\right);$$

Hol. square. $A = h^2 - h'^2, \quad x_0 = \frac{h}{6}\left(1 + \frac{h'^2}{h^2}\right);$

Hol. circle. $A = \pi(r^2 - r'^2), \quad x_0 = \frac{d}{8}\left(1 + \frac{r'^2}{r^2}\right).$

The ultimate compressive strengths of various rocks used in foundations are approximately, for

| | | |
|---|---|---|
| Granite................ | 12800 | pounds per square inch. |
| Sandstone............. | 9800 | " " " " |
| Soft sandstone......... | 3000 | " " " " |
| Strong limestone....... | 8500 | " " " " |
| Weak limestone........ | 3000 | " " " " |
| Hard red brick......... | 3000 | " " " " |
| Common brick......... | 1000 | " " " " |
| Portland cement concrete: | | |
|   1 month old.......... | 1000 | " " " " |
|   12 months "........... | 6000 | " " " " |
| Rosendale cement concrete: | | |
|   6 months old.......... | 1200 | " " " " |

## FOUNDATIONS UPON EARTH.

*Firm Earth.*—Earth which has an angle of repose of at least 27° may be considered as firm, and for foundation purposes requires little preparation other than the excavation of a trench or pit, and making the surface receiving the masonry level. From Table II it is seen that sand, **gravel,** and damp clay are classed as firm soils; however, these may become so saturated with water that their angles of repose will become considerably less than 27°, hence precautions must be taken against too much water by draining the ground in the immediate vicinity of the foundation. Particular care must be taken in the case of clay, or sand which will become a kind of quicksand when saturated with water.

Before attempting to design a foundation, the character of the earth must be determined either by test excavations, borings, or from the experience of others. It often happens that from all surface indications the earth appears to be firm, but upon excavating it is found there is a stratum of semi-fluid mud or quicksand underneath; in such cases care must be taken to determine the minimum thickness of the stratum of firm earth, for if too thin it will not be safe to build upon, and then a foundation has to be prepared according to some of the methods described later.

Considering the earth as a homogeneous granular mass, the supporting power at any depth can be computed when the angle of repose $\phi$ is known. Some practical men object to any theoretical formulas being employed in connection with the determination of the bearing or supporting power of earth, claiming that the assumptions upon which the formulas are based are rarely if ever found in practice. This is probably true to a certain extent, yet the theoretical formulas are upon the safe side, and do not lead to absurd results; in fact, the results obtained by their judicious application agree very well with the practice of the best engineers.

If $p =$ the maximum supporting power per square foot of earth;

$\gamma =$ the weight of one cubic foot of earth;

$\phi =$ the angle of repose;

and $x'$ = the depth of the plane below the surface upon which the maximum supporting **power** is desired;

then

$$p = x'\gamma \left\{ \frac{1 + \sin \phi}{1 - \sin \phi} \right\}^2 \text{ (see page 40)}. \quad . \quad . \quad (1)$$

And if $p'$ is the **minimum** intensity of the pressure upon the earth which is **allowable for the** stability of the earth and the foundation with its load,

$$p' = x''\gamma \left\{ \frac{1 - \sin \phi}{1 + \sin \phi} \right\}^2 \text{ (see page 40)}, \quad . \quad . \quad (2)$$

where $x''$ is the **depth of the plane** considered below the surface of the earth.

The above equations **neglect any friction** between the earth and the masonry of the foundation. In deep foundations this is a large factor on the safe side.

If the surface of the earth is level, then $x' = x''$; and further, if the earth is subjected to a uniformly distributed load only the average intensity need be considered.

Equation (2) is considerably different from that given by Rankine, and writers who have followed him, in this, that they consider the minimum intensity allowable to be equal to $x''\gamma$ = the **average intensity** of the pressure upon a plane at a depth $x''$ in an unlimited mass. This does not appear to the writer to be a **logical** treatment of the subject, if the mass has an **angle of repose greater** than zero, and the maximum intensity allowable be determined as a function of this angle.

According to the assumption of Rankine, it would appear that if a box without a bottom were **sunk** into a mass of perfectly dry sand it would be filled from the bottom until the surfaces without and within were at the same level; but this does not take place, and would not even if the sides of the box were frictionless. The sand only fills the box partially, or until the requirements of equation (2) are fulfilled. Hence it seems to the writer that if the maximum intensity is a function of $\phi$, the value of the minimum intensity must be also.

From equations (1) and (2) it is evident that the allowable intensity upon the earth of any pressure or load commonly called the supporting power varies *directly as the depth*, as long as $\phi$ remains unchanged; hence all tables of supporting powers of earth are of little value unless the depth of the foundation upon which they are based is known. Unfortunately this is omitted in most cases, and only the character of the earth is given. The depth to which foundations must be sunk in many localities has a *minimum* value governed by the depth to which frost extends. This is not always true, however, as in Terre Haute, Indiana, frame houses and brick blocks two and one-half stories high are constructed practically upon the surface, the sod only being removed. The width of the foundation is not excessive, but on the contrary narrow. No serious settlement results, owing to the character of the earth, which is very sandy, and will not retain sufficient moisture to permit frost action to heave the structures. The actual load per square foot supported by the soil is about one ton. If $x'$ be taken as one foot, $\gamma$ as 100 pounds, and $p$ as 2000 pounds, then from equation (1) $\phi$ is about 39°, which is below the actual value.

The above case, however, may be called an exception to the general rule that all foundations must be sunk below the action of frost, or to a depth of three feet or more according to the locality.

For convenience the values of

$$\left(\frac{1 + \sin \phi}{1 - \sin \phi}\right)^2 \quad \text{and} \quad \left(\frac{1 - \sin \phi}{1 + \sin \phi}\right)^2$$

are given in the following table:

| $\phi$ | $\left(\frac{1+\sin\phi}{1-\sin\phi}\right)^2$ | $\left(\frac{1-\sin\phi}{1+\sin\phi}\right)^2$ | $\phi$ | $\left(\frac{1+\sin\phi}{1-\sin\phi}\right)^2$ | $\left(\frac{1-\sin\phi}{1+\sin\phi}\right)^2$ |
|---|---|---|---|---|---|
| 0  | 1.00 | 1.00 | 23 | 5.21  | 0.19 |
| 5  | 1.42 | 0.70 | 24 | 5.62  | 0.18 |
| 6  | 1.52 | 0.66 | 25 | 6.07  | 0.16 |
| 7  | 1.63 | 0.61 | 26 | 6.56  | 0.15 |
| 8  | 1.75 | 0.57 | 27 | 7.09  | 0.14 |
| 9  | 1.88 | 0.53 | 28 | 7.67  | 0.13 |
| 10 | 2.02 | 0.50 | 29 | 8.30  | 0.12 |
| 11 | 2.16 | 0.46 | 30 | 9.00  | 0.11 |
| 12 | 2.32 | 0.43 | 31 | 9.76  | 0.10 |
| 13 | 2.50 | 0.40 | 32 | 10.59 | 0.09 |
| 14 | 2.68 | 0.37 | 33 | 11.50 | 0.09 |
| 15 | 2.88 | 0.35 | 34 | 12.51 | 0.08 |
| 16 | 3.10 | 0.32 | 35 | 13.62 | 0.07 |
| 17 | 3.33 | 0.30 | 36 | 14.84 | 0.07 |
| 18 | 3.59 | 0.28 | 37 | 16.18 | 0.06 |
| 19 | 3.86 | 0.26 | 38 | 17.67 | 0.06 |
| 20 | 4.22 | 0.24 | 39 | 19.64 | 0.05 |
| 21 | 4.48 | 0.22 | 40 | 21.16 | 0.05 |
| 22 | 4.83 | 0.21 |    |       |      |

Having determined upon the **depth** to which it is expedient to extend the foundation, a *minimum* value of $\phi$ must be assumed from a knowledge of the earth, and then the **allowable** bearing or supporting power can be found from equations (1) and (2); **or if** the supporting power is assumed, the minimum depth to which the foundation must be sunk can **be found from** the same equations.

The **proper** proportions of the foundation are most easily obtained from the following equations, which are deduced for a few of the ordinary forms and conditions. All masonry foundations are usually trapezoidal in section, and hence formulas based upon this form can be applied to stepped foundations without serious error.

CASE I. *Given a uniformly distributed load to be supported by symmetrical trapezoidal foundation sunk to a known depth, to determine the minimum width of the base of the foundation.*

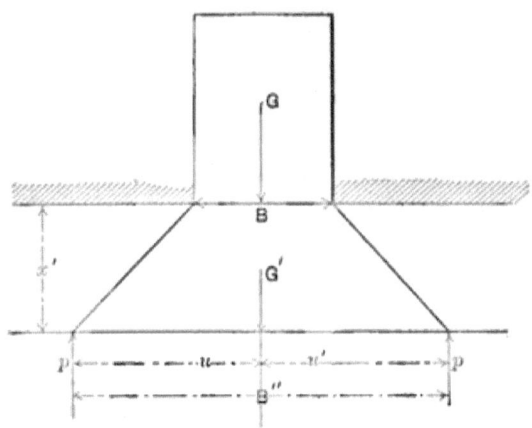

FIG. 85.
Section of Wall and Foundation.

Let $G =$ the total weight to be supported less that of the foundation;

$G'' = G +$ weight of the foundation;

and $B'' =$ minimum breadth of the foundation.

Assuming $x'$, the value of $p$ is

$$p = x'\gamma \left\{ \frac{1 + \sin \phi}{1 - \sin \phi} \right\}^2.$$

From the figure

$$G' = G + W \frac{B + B''}{2} x' = B''p;$$

or
$$B'' = \frac{2G + BWx'}{2p - Wx'}.$$

*The above formula applies to a wall one foot long.*—In case of an isolated pier, the value of $x'$ can be found as above. $B''$ may be assumed and a rough calculation made to determine if the average pressure upon the earth is equal to or less than $p$. A second trial usually determines the proper value for $B''$. The exact formula for the determination of the dimensions of a square or rectangular foundation with stepped sides is an equation of the second degree.

Ex. 1. A trapezoidal foundation 5 feet broad on top has to support 50,000 pounds per lineal foot in length, in earth having a minimum angle of repose of 30°. The maximum depth to which the foundation is to be sunk is 5 feet; determine $B''$ and $p$, when $\gamma = 100$ pounds and $W = 150$ pounds.

From (1)
$$p = 5 \cdot 100 \cdot 9 = 4500 \text{ pounds—say } 4000;$$
then
$$B'' = \frac{100000 + 3750}{8000 - 750} = 14.3;$$

or the proper width of the base is about 14.5 feet.

Ex. 2. A cast-iron plate, 2 feet square under a column, transmits a load of 20,000 pounds to a masonry foundation 3 feet square. How deep must this be sunk in earth when $\phi = 30°$, $\gamma = 100$ pounds, and $W = 150$ pounds?

Neglecting the weight of the masonry in the foundation, the intensity of the pressure upon the earth is about 2200 pounds; then from (1) $x' =$ about 2.5 feet—say 3 feet.

The actual intensity of the pressure upon the earth is now $\dfrac{20{,}000 + 4050}{9} = 2670$ pounds.  Substituting this value of $p$ in (1) and solving for $x'$, its value is 2.85 feet; hence 3 feet is the required depth of the foundation.

The weight of the earth supported by the masonry of the foundation is neglected.

Case II. *Unsymmetrical distribution of pressure upon the base of a foundation.*

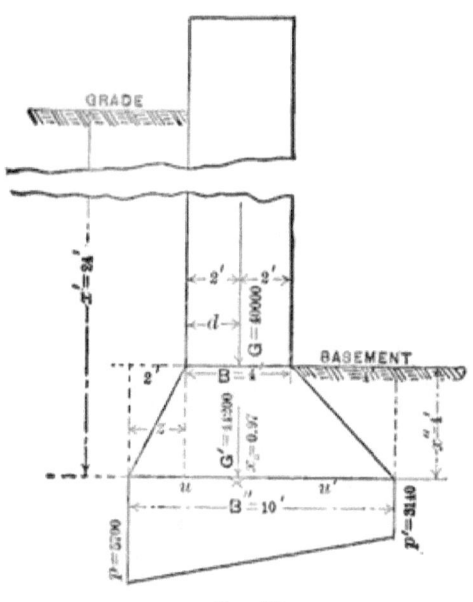

Fig. 36.
Section of Wall and Foundation.

One of the many examples of pressure unevenly distributed upon the bed of a foundation is the case of an outside wall of a building located very near the property line and circumstances prevent encroaching upon the neighboring property to any great extent. Here two conditions must be fulfilled. The maximum intensity of the pressure $p$, Fig. 36, must not be greater than the supporting power of the earth at the depth $x'$, and the minimum intensity $p'$ must not be so small that the earth having a depth $x''$ may tend to heave the foundation.

Let $p_o$ = the *average* intensity of the pressure upon the base. Then

$$p_o = \frac{G'}{B''} = \frac{p+p'}{2}, \quad p = x'\gamma\left\{\frac{1+\sin\phi}{1-\sin\phi}\right\}^2.$$

But

$$G' = G + \frac{B+B''}{2}x''W.$$

Therefore

$$B'' = \frac{2G + BWx''}{2p_o - Wx''};$$

in which $x''$ is determined from the equation

$$p' = x''\gamma\left\{\frac{1-\sin\phi}{1+\sin\phi}\right\}^2.$$

It is thus possible to determine $B''$ quite easily, but the value of the offset $z$ so that $p$ and $p'$ shall have their proper values must be either found by trial or computation. Since one or two trials are sufficient to determine $z$, the formula will not be given here.

Ex. 3. In Fig. 36, page 75, let $G = 40{,}000$ pounds, $B = 4$ feet, $d = 2$ feet, $x' = 24$ feet, and $x'' = 4$ feet. If the thrust of the earth be neglected, what must be the width of the base of the foundation, so that the average pressure per unit area shall not exceed 4800 pounds, and the maximum 7000 pounds, when $\gamma = 100$, $W = 150$,

$\phi = 30°$? The bulk of the foundation to be on the right of the centre of the wall.

First determine the allowable intensities,

$$\begin{aligned}
\max p &= x'\gamma(9) &&= 2400 \times 9 &&= 21600 \text{ pounds.} \\
\min &= x'\gamma(0.11) &&= 2400 \times 0.11 &&= 264 \text{ ``} \\
\max p' &= x''\gamma(9) &&= 400 \times 9 &&= 3600 \text{ ``} \\
\min &= x''\gamma(0.11) &&= 400 \times 0.11 &&= 44 \text{ ``}
\end{aligned}$$

From the formula on page 76

$$B'' = \frac{2G + BWx''}{2p_0 - Wx''} = \frac{82400}{9000} = 9.15 \text{ feet.}$$

Take 10 feet as the value of $B''$; then the weight of the masonry in the foundation is 4200 pounds, and

$$p_0 = \frac{44200}{10} = 4420.$$

By graphics or by moments, assuming $z = 2$ feet, the resultant pressure cuts the base 0.97 foot from the centre, and hence $p = 5700$ pounds and $p' = 3140$ pounds.

The above width of base and the intensities just obtained satisfy all the conditions of the problem, though the value of $z$ could be decreased a little, increasing the intensity at the toe and decreasing that at the heel.

*Projection of Footing-courses.*—Where masonry foundations are stepped as is the usual custom, the proper offset for each course may be determined as follows, by considering each offset as a cantilevered beam of stone *uniformly loaded:*

Let $o$ = the offset of any particular course;

$p_0$ = the intensity of the pressure upon the base of the course;

$t$ = the thickness of the course;
$R$ = the modulus of rupture of the material; and
$F$ = the factor of safety.

Then
$$p_\circ \frac{o^2}{2} = \frac{1}{6} \frac{R}{F} t^2,$$

or

$$o = t \sqrt{\frac{R}{F} \frac{1}{3 p_\circ}}.$$

In case the intensity of the pressure is not uniform, but varies uniformly from one side to the other, the quantity $p_\circ$ may be replaced by $p$, the maximum intensity for the offset on the side having the greater pressure, and by $p'$, the minimum intensity for the steps or offsets on the side of the lesser pressure : in the first case the factor of safety will be slightly increased and in the second decreased.

The above formula is applicable only when the stones project less than half their length and when thoroughly well laid in cement mortar.

The table on the following page is given by Prof. Baker.

Other factors remaining the same, the offsets vary *directly* as the square roots of the moduli of rupture and inversely as the factors of safety, so that the above table can be applied for any values of $R$ and $F$ by simple proportion.

*Foundations upon Soft Earth.*—When a foundation must be placed upon soft earth which offers no particular difficulties other than the requirement of broadness or depth of the excavation, considerable expense can be avoided by excavating the soft material and replacing it by firm material, or by driving short piles spaced about three feet on centres, commencing at the outer limits of the foundation and working towards the centre, and thus compressing the earth; sometimes holes are bored and filled with sand, making sand-piles, etc. The proper depth and spread of such foundations can be found from formulas (1) and (2) by including the prepared earth as a portion of the foundation.

SAFE OFFSETS FOR MASONRY FOOTING-COURSES,
IN TERMS OF THE THICKNESS OF THE COURSE, USING 10 AS A FACTOR OF SAFETY.

| Kind of Stone. | $R$ in Lbs. per Sq. In. | Offsets for a Pressure, in Tons per Sq. Ft., on the Bottom of the Course of Masonry. | | |
|---|---|---|---|---|
| | | 0.5 | 1.0 | 1.5 |
| Bluestone flagging .................. | 2700 | 3.6 | 2.6 | 1.8 |
| Granite............................. | 1800 | 2.9 | 2.1 | 1.5 |
| Limestone.......................... | 1500 | 2.7 | 1.9 | 1.3 |
| Sandstone .......................... | 1200 | 2.6 | 1.8 | 1.3 |
| Slate............................... | 5400 | 5.0 | 3.6 | 2.5 |
| Best hard brick ................... | 1500 | 2.7 | 1.9 | 1.3 |
| Hard brick......................... | 800 | 1.9 | 1.4 | 0.8 |
| Concrete { 1 Portland, 2 sand, 3 pebbles } 10 days old. | 150 | 0.8 | 0.6 | 0.4 |
| Concrete { 1 Rosendale, 2 sand, 3 pebbles } 10 days old. | 80 | 0.6 | 0.4 | 0.3 |

In case the earth has sufficient water to keep the foundation damp, a very excellent foundation upon soft earth is a platform of timber composed of heavy sticks laid close together in layers, every alternate layer being right-angled with that adjacent, and thoroughly driftbolted together. Another method is to form a grillage of the timbers and fill the spaces around the sticks with concrete.

In dry soft earth the timber platform may be replaced by a bed of concrete, which is more durable, but not as elastic. Recently the combination of iron or steel beams with concrete has been successfully employed for foundations upon soft earth in Chicago.

The safe projection of the timber platform or one of concrete beyond the masonry can be found by the formula already given.

The safe projection of iron or steel beams can be found as follows:

Let $I =$ the moment of inertia of the section;
$h =$ the depth of the beam;
$p_o =$ the intensity of the pressure upon the bed of the foundation transmitted to the beam;
$R =$ the modulus of rupture of the material composing the beam;
and $F =$ the factor of safety.
Then

$$p_o \frac{o^2}{2} = 2\frac{RI}{Fh}$$

or

$$o = 2\sqrt{\frac{R}{F}\frac{1}{p_o h}I}$$

In case the pressure upon the base of the foundation is not uniform, the method outlined for masonry offsets can be applied in proportioning the offsets of steel or iron beams.

The following table of the safe projections of steel I beams is given in Carnegie's Pocket Companion.

FOUNDATIONS UPON EARTH. 81

## TABLE

GIVING SAFE LENGTHS OF PROJECTIONS "$o$" IN FEET (SEE ILLUSTRATION), FOR "$s$" $= 1$ FOOT AND VALUES OF "$p_0$" RANGING FROM 1 TO 5 TONS.

| Depth of Beam, in. | Weight per Foot, lbs. | $b$ (Tons per Square Foot.) | | | | | | | | | |
|---|---|---|---|---|---|---|---|---|---|---|---|
| | | 1 | $1\frac{1}{4}$ | $1\frac{1}{2}$ | 2 | $2\frac{1}{4}$ | $2\frac{1}{2}$ | 3 | $3\frac{1}{2}$ | 4 | $4\frac{1}{2}$ | 5 |
| 20 | 80 | 14.0 | 12.5 | 11.5 | 10.0 | 9.0 | 9.0 | 8.0 | 7.5 | 7.0 | 6.5 | 6.0 |
| 20 | 64 | 12.5 | 11.0 | 10.0 | 8.5 | 8.0 | 8.0 | 7.0 | 6.5 | 6.0 | 6.0 | 5.5 |
| 15 | 80 | 12.0 | 10.5 | 9.5 | 8.5 | 8.0 | 7.5 | 7.0 | 6.5 | 6.0 | 5.5 | 5.0 |
| 15 | 60 | 10.5 | 9.5 | 8.5 | 7.5 | 7.0 | 6.5 | 6.0 | 5.5 | 5.5 | 5.0 | 5.0 |
| 15 | 50 | 9.5 | 8.5 | 8.0 | 7.0 | 6.5 | 6.0 | 5.5 | 5.0 | 5.0 | 4.5 | 4.5 |
| 15 | 41 | 8.5 | 8.0 | 7.0 | 6.0 | 6.0 | 5.5 | 5.0 | 4.5 | 4.5 | 4.0 | 4.0 |
| 12 | 40 | 8.0 | 7.0 | 6.5 | 5.5 | 5.5 | 5.0 | 4.5 | 4.0 | 4.0 | 3.5 | 3.5 |
| 12 | 32 | 7.0 | 6.5 | 5.5 | 5.0 | 4.5 | 4.5 | 4.0 | 4.0 | 3.5 | 3.5 | 3.0 |
| 10 | 33 | 6.5 | 6.0 | 5.5 | 4.5 | 4.5 | 4.0 | 4.0 | 3.5 | 3.5 | 3.0 | 3.0 |
| 10 | 25.5 | 5.5 | 5.0 | 4.5 | 4.0 | 4.0 | 3.5 | 3.5 | 3.0 | 3.0 | 2.5 | 2.5 |
| 9 | 27 | 5.5 | 5.0 | 4.5 | 4.0 | 4.0 | 3.5 | 3.5 | 3.0 | 3.0 | 2.5 | 2.5 |
| 9 | 21 | 5.0 | 4.5 | 4.0 | 3.5 | 3.5 | 3.0 | 3.0 | 2.5 | 2.5 | 2.5 | 2.0 |
| 8 | 22 | 5.0 | 4.5 | 4.0 | 3.5 | 3.5 | 3.0 | 3.0 | 2.5 | 2.5 | 2.5 | 2.0 |
| 8 | 18 | 4.5 | 4.0 | 3.5 | 3.0 | 3.0 | 3.0 | 2.5 | 2.5 | 2.0 | 2.0 | 2.0 |
| 7 | 20 | 4.5 | 4.0 | 3.5 | 3.0 | 3.0 | 3.0 | 2.5 | 2.5 | 2.0 | 2.0 | 2.0 |
| 7 | 15.5 | 4.0 | 3.5 | 3.0 | 2.5 | 2.5 | 2.5 | 2.0 | 2.0 | 2.0 | 2.0 | 1.5 |
| 6 | 16 | 3.5 | 3.0 | 3.0 | 2.5 | 2.5 | 2.0 | 2.0 | 2.0 | 1.5 | 1.5 | 1.5 |
| 6 | 13 | 3.0 | 3.0 | 2.5 | 2.5 | 2.0 | 2.0 | 2.0 | 1.5 | 1.5 | 1.5 | 1.5 |
| 5 | 13 | 3.0 | 2.5 | 2.5 | 2.0 | 2.0 | 2.0 | 1.5 | 1.5 | 1.5 | 1.5 | 1.5 |
| 5 | 10 | 2.5 | 2.5 | 2.0 | 2.0 | 1.5 | 1.5 | 1.5 | 1.5 | 1.5 | .. | .. |
| 4 | 10 | 2.5 | 2.0 | 2.0 | 1.5 | 1.5 | 1.5 | 1.5 | .. | .. | .. | .. |
| 4 | 7.5 | 2.0 | 2.0 | 1.5 | 1.5 | 1.5 | 1.5 | .. | .. | .. | .. | .. |

Above table applies to *steel* beams. Values given based on extreme fibre stresses of **16,000 pounds per square inch**.

*Pile Foundation.*—Pile foundations are employed in all kinds of earth, sometimes to save expense and sometimes because nothing else appears to be as good. In localities where the earth is uncertain in its character the use of piles enables the engineer to put in a foundation which he feels sure is safe, as a single pile thirty feet long will support several tons even when driven into mud, the load in this case being carried almost entirely by the friction of the

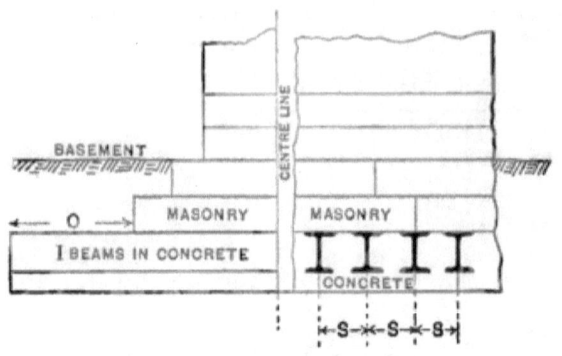

Fig. 37.

mud upon the surface of the pile. If the pile is driven through the mud to a solid stratum below, then the pile acts as a column more or less supported its entire length, and consequently able to carry a very great load.

Piles are usually spaced about three feet on centres, and the tops firmly bedded in a layer of concrete or stayed by a grillage of timber or by a combination of these methods, the object being to thoroughly and evenly distribute the load to be supported.

The supporting power of a pile in a given earth can be found in the following manner:

Let $G'$ = the total load to be supported by the pile, including the weight of the pile;

$p_0$ = the intensity of the pressure upon the bottom of the pile;

$A$ = the superficial area of the pile in contact with the earth;

and $f =$ a factor depending upon the friction resistance of a unit area of the surface of the pile.

Then for a pile having a diameter of $d$

$$* \; G' = \frac{\pi d^2}{4} p_\circ + fA.$$

But

$$p_\circ = x'\gamma \left\{ \frac{1 + \sin \phi}{1 - \sin \phi} \right\}^2 \quad \text{and} \quad A = \pi d x'.$$

$$\therefore \; x' = \frac{G'}{\gamma \dfrac{\pi d^2}{4} \left\{ \dfrac{1 + \sin \phi}{1 - \sin \phi} \right\}^2 + f\pi d}.$$

For practical purposes this may be written

$$x' = \frac{G'}{\gamma \left\{ \dfrac{1 + \sin \phi}{1 - \sin \phi} \right\}^2 + 3f}.$$

For convenience this may be further simplified for special cases.

The following values of $f$ have recently been given by W. M. Patton, based upon his own and the experience of others:

In very soft silt or liquid mud, $f = 150$ pounds per sq. ft.
In ordinary clay or earth (dry), $f = 300$ " " " "
"  "  "  "  " (wet), $f = 150$ " " " "
In compact hard clay, $f = 300$ " " " "
In sand, or sand and gravel, $f = 500$ " " " "

---

* This formula was suggested by reading W. M. Patton's article on piles in his "Practical Foundations,"

*For the silt of swamps, muds, etc.*, φ is very nearly if

### ERRATA.

On page 85, formulæ should read as follows:

$$G' = \frac{\sqrt[3]{h} \times W \times 0.023}{2(1+a)} \quad \text{(Trautwine's)};$$

$$G' = \frac{2wh}{a+1} \quad (\text{Eng. News});$$

where
- $G'$ = the safe load in *tons*;
- $W$ = the weight of the hammer in *pounds*;
- $w$ =   "       "      "      "     "   *tons*;
- $h$ = the fall of the hammer in *feet*;

and
- $a$ = the average penetration of the pile in *inches* during the last few blows.

be driven can be determined by borings, and thus φ and γ determined upon.

The value of $f$ can be found by studying the behavior of piles already driven in similar earth. Thus it appears that the above formula must be as accurate in results and as safe in application as the majority of the formulas used by engineers in proportioning structures.

The formula is independent of the means by which the pile is driven, as ought to be the case, since very often piles are sunk by water-jets, or even by working them backwards and forward, making the formulas depending upon the weight of a driving-hammer, its fall, and the penetration of the pile during the last few blows useless. Two of the most simple of the many formulas of this class are those of Trautwine and the *Engineering News*, viz.:

$$G' = \frac{3\sqrt{h} \times W \times 0.0268}{2(1+a)} \quad \text{(Trautwine's);}$$

$$G' = \frac{2Wh}{a+1} \quad (Eng.\ News);$$

where  $G'$ = the safe load;
$W$ = the weight of the hammer in pounds;
$h$ = the fall of the hammer in inches;
and  $a$ = the average penetration of the pile in inches during the last few blows.

*Screw-pile.*—Screw-piles are usually round, and have at the bottom a cast or wrought iron screw. The piles are of wood, cast iron, or wrought iron. The diameter of the screw is from two to eight times the diameter of the pile, and its pitch from one fourth to one half its diameter. The screw seldom has but one turn. The piles are sunk by turning them by means of levers or by power. (Fig. 45.)

The load which the pile will carry depends principally upon the supporting power of the earth at the depth of the screw and the area of the screw, though in all cases there is more or less frictional resistance upon the surface of the pile proper. If $x'$ is the depth of the screw and $p_0$ the allowable intensity of the pressure upon the earth at that depth, then

$$p_o = \frac{x'}{\gamma} \left\{ \frac{1 - \sin \phi}{1 + \sin \phi} \right\}^2.$$

Screw-piles can be advantageously employed for supporting structures above water where the upper ends of the piles can be used as columns. They are chiefly employed in light-house construction.

*Sheet-piles.*—Sheet-piles are usually of wood in the form of planks, and are driven as closely together, edge to edge, as possible, the object being to form a water-tight barrier.

To make the joints tight the planks are oftened tongued and grooved. A patent sheet-pile is formed by bolting together three planks of equal width, so that the middle plank will form the tongue on one side and the outside planks the groove on the other side. Sheet-piles are also employed to confine soft earths.

# FOUNDATIONS UNDER WATER AND DEEP FOUNDATIONS.

Foundations under water differ in general but little from those upon dry earth, the effect of water, ice, etc., upon the structure, however, constitute additional problems to be solved for each locality.

A few of the various methods employed in placing foundations under water or at great depths will be very briefly described.

*Coffer-dams.*—A coffer-dam is merely a tight wall surrounding the locality where the foundation is to be placed, excluding water from the enclosure, which can be pumped dry and the surface prepared to receive the foundation.

In quiet and shallow water the dam may be made of earth, or sheet-piles banked with earth.

In deep water large piles are driven every few feet in two rows around the site, to which horizontal timbers are bolted, acting as guides and supports to a double row of sheet-piles, between which is placed puddled earth. To prevent bending, the large piles are cross-tied with bolts.

The space enclosed should be somewhat larger than required by the foundation, to allow room for materials, etc. (Fig. 46.)

*Timber Cribs.*—A timber crib is a box built of large timbers and divided into cells by cross partitions. The joints and splices of the timbers employed are arranged so that walls and partitions are thoroughly tied together. In case a tight wall-crib is wanted the timbers may be dapped one fourth their depth on both sides or halved together. Cribs are built in the shape best suited to the purpose for which they are to be used. They are usually constructed at some convenient point near the site of the foundation, and then towed to the place where they are to be sunk. In constructing the crib a few of the cells are planked near the bottom. These are filled with stone until the crib sinks to the surface previously prepared to receive it. The other cells are now filled with stone and the regular masonry commenced. Sometimes the top of the crib is planked over before the masonry is started. (Fig. 44.)

The surface which is to receive the crib may be soft mud, riprap, rock, or piles. The crib is allowed to sink into the mud and to rest upon riprap which has been levelled. If the surface is level rock, the crib is merely sunk; but if the rock is uneven, it is either levelled or the crib is sunk until it just touches rock at some point, when riprap is thrown around and under the crib.

Timber cribs are extensively employed in various classes of engineering works for both temporary and permanent structures.

In permanent structures the timbers supporting masonry, etc., should always be under water.

Timber cribs are sometimes used as coffer-dams by making the outside cells water-tight. The crib is sunk into the mud, or the bottom edges banked with earth, etc., until the interior can be kept dry by pumping.

*Open Caissons.*—An open caisson is a strong water-tight box which is floated to the site of the foundation and sunk to its place by the masonry proper, which is built inside the box. After the bottom has reached its position and the top of the masonry is above water, the sides are removed, leaving the bottom of the box as a platform supporting the masonry. The surface to receive an open caisson is prepared by dredging, throwing in riprap, driving piles, etc., as best suits the locality. (Fig. 47.)

*Cushing Cylinder Piers.*—A cluster of piles is first driven as closely together as possible, and their tops thoroughly bolted one to the other. Then an iron cylinder is placed around the cluster and built up in sections until the top is above water. Then the cylinder is made to sink by dredging out the material inside by water-jets, by disturbing the material around the edges, etc., until a desired depth is reached, sections being bolted to the top of the cylinder as needed. The cylinder is now filled with concrete to the top and covered with an iron cap which receives the load to be carried. The size and number of cylinders employed depends upon the superstructure.

For ordinary bridges two cylinders cross-braced form a pier.

The supporting power depends upon the piles principally, though the friction upon the outside of the cylinders offers some resistance to settlement.

*Pneumatic Caissons.*—A pneumatic caisson is essentially an air-tight box with the open side imbedded in earth, from which the air is pumped to allow the box to sink or into which air is pumped to prevent sinking. In water the caisson usually carries a water-tight timber crib, which in turn supports a timber coffer-dam, the crib enabling the structure to be loaded with stone according to the requirements of the sinking operation, and the coffer-dam keeping the water out near the surface. Various combinations of caisson, crib, and coffer-dam are made, however, to suit conditions. (Fig. 48.)

The ordinary method of sinking caissons is to pump in

enough air to exclude water from the chamber, while laborers dig out the material over the surface and near the edges of the chamber, this material being removed by various methods such as pumps, lifts, etc. When sufficient material has been removed, all the laborers leave the caisson, leaving one man only who watches for leaks; the air-pressure is then lowered a little, and the caisson with its superstructure sinks. This process is repeated until a solid foundation is reached, when the caisson is filled with concrete, as also are the cribs, etc., if any, above the caisson.

## Types of Existing Foundations.

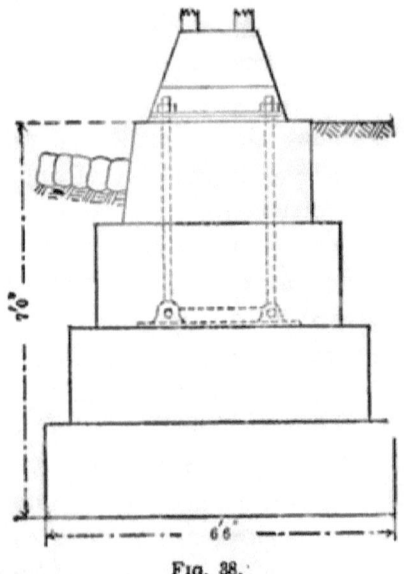

Fig. 38.

Concrete Pier used as Foundation for Elevated Railroad Columns (*Engineering and Building Record*, Sept. 14, 1895.)

TYPES OF EXISTING FOUNDATIONS. 91

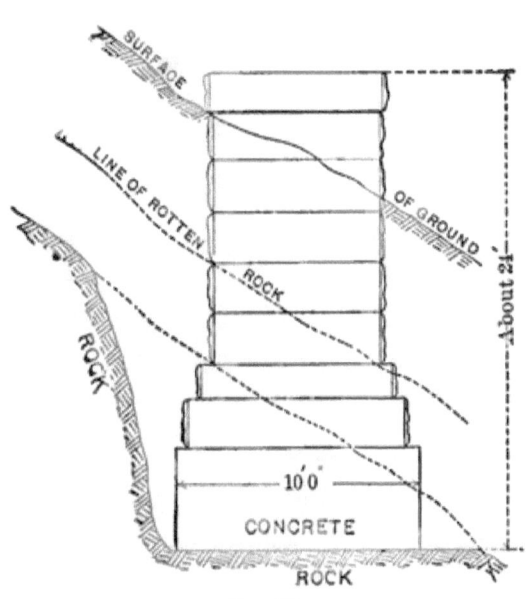

FIG. 39.

Elevation of Masonry Pier with Bottom Course of Concrete. Illustrating the removal of rotten rock and the levelling of the rock surface. (Marent Gulch Viaduct, N. P. R. R.; *Trans. Am. Soc. C. E.*, Sept., 1891.)

## 92 TYPES OF EXISTING FOUNDATIONS.

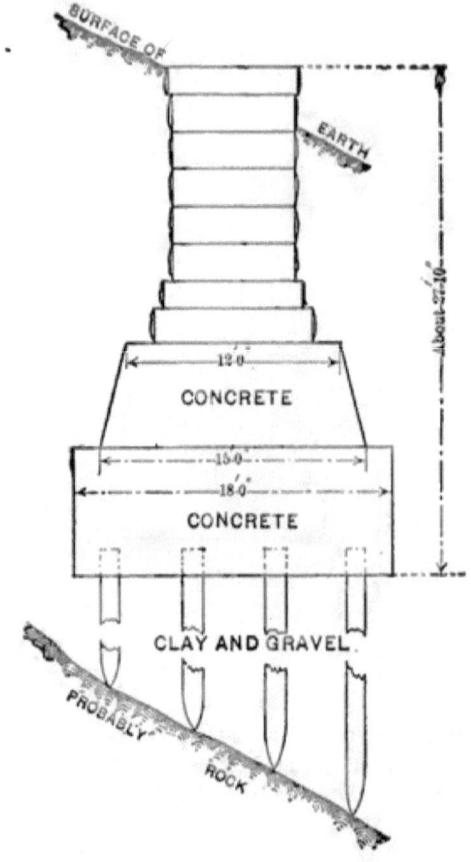

FIG. 40.

Elevation of another Pier of the Marent Viaduct Foundations. Showing the application of piles and concrete to obtain a solid foundation.

## TYPES OF EXISTING FOUNDATIONS. 93

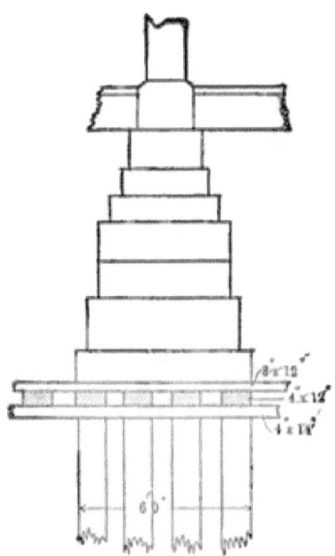

FIG. 41.

Elevation of a Pier in the Foundation of a Chicago Grain Elevator. Illustrating the use of piles and a wooden platform in soft ground. Piles are from 20 to 40 feet long, and reach hardpan. Twelve piles are placed under each post, and each pile supports a load of about 22 tons. (*Engineering and Building Record*, Nov. 12, 1895.)

94     TYPES OF EXISTING FOUNDATIONS.

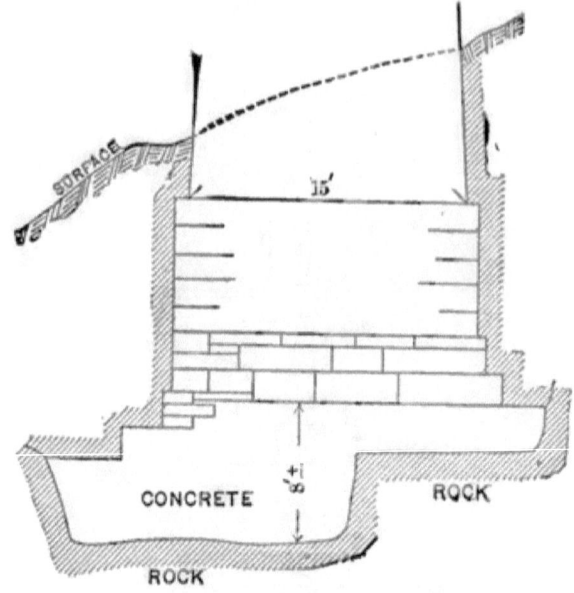

Fig. 42.

End Elevation of Masonry Pier supporting Stone Arches of Washington Bridge. Illustrating the use of concrete to level the rock surface to receive masonry.

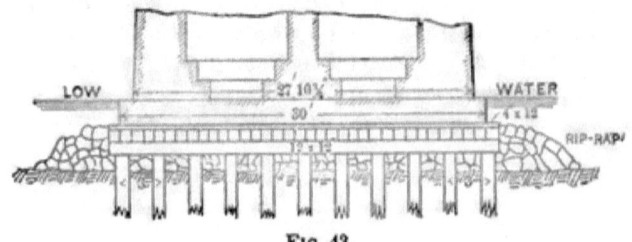

Fig. 43.

Section through Centre of Foundation of Pivot Pier of Grand Forks Bridge. Illustrating the use of piles, wooden platform, and rip-rap. (*Baker.*)

## TYPES OF EXISTING FOUNDATIONS. 95

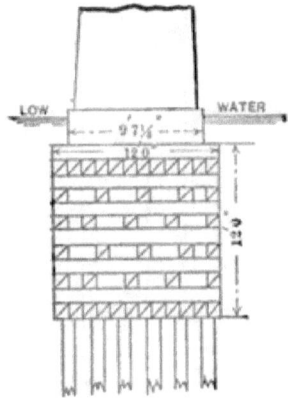

FIG. 44.

End Elevation of Foundation of Pier of Croix River Bridge. Illustrating the use of timber crib and piles. (*Baker*.)

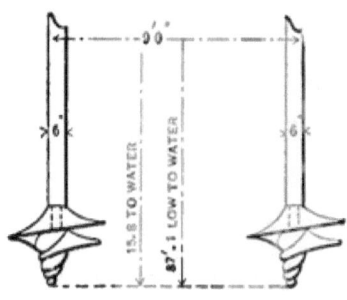

FIG. 45.

Mobile River Bridge Piers. Composed of two rows of screw-piles, about 9 feet centre to centre, with piles spaced about 8 feet apart. (See *Engineering News*, vol. xiii. p. 210.)

96    TYPES OF EXISTING FOUNDATIONS.

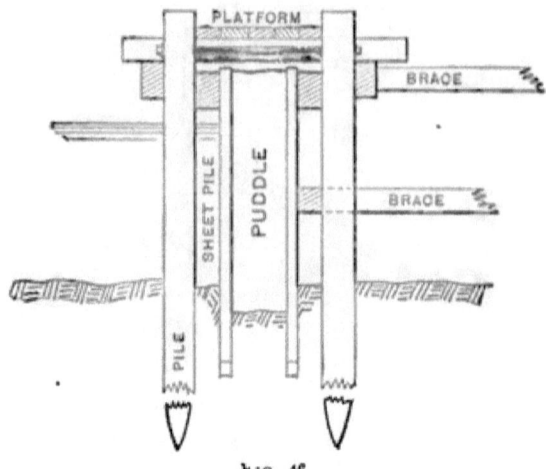

Fig. 46.
Sketch showing Cross-section of Coffer-dam.

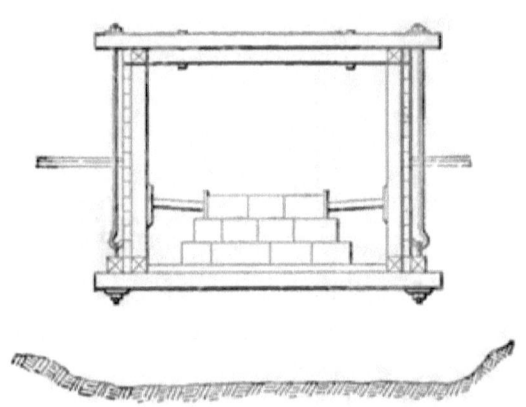

Fig. 47.
Sketch showing Essential Features of Open Caisson.

## TYPES OF EXISTING FOUNDATIONS.

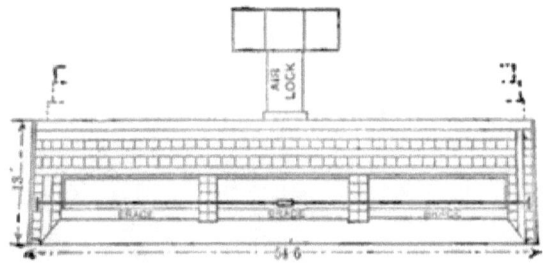

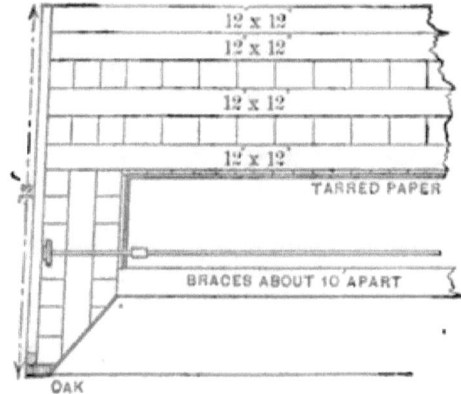

Fig. 48.

Section of One of the Caissons employed in the Foundations of the Piers for the Washington Bridge.

# REFERENCES.

## EARTH-PRESSURE AND RETAINING-WALLS.

A brief outline of the theories advanced by the following writers can be found in "*Neue Theorie des Erddruckes*," Dr. E. Winkler, *Wien*, 1872:

| | | |
|---|---|---|
| D'Antony, | Hoffmann, | Poncelet, |
| Andé, | Holzhey, | Prony, |
| Andoy, | de Lafont, | Rankine, |
| Belidor, | Levi, | Rebhann, |
| Blaveau, | de Köszegh Martony, | Rondelet, |
| Bullet, | Maschek, | Saint-Guilhem, |
| Considère, | Mayniel, | Saint-Venant, |
| Coulomb, | Mohr, | Sallonnier, |
| Couplet, | Montlong, | Scheffler, |
| Culmann, | Moseley, | Trincaux, |
| Français, | Navier, | Vauban, |
| Gadroy, | Ortmann, | Winkler, |
| Gauthey, | v. Ott, | Woltmann. |
| Hagen, | Persy, | |

AUDÉ. Poussée des Terres. Nouvelles expériences sur la poussée des terres. Paris, 1849.

BAKER-CURIE. Note sur la brochure de M. B. Baker théorie. Annales des Ponts et Chaussées, pp. 558–592, 1882.

—— The actual lateral pressure of earthwork. Van Nostrand's Magazine, xxv, 1881; also Van Nostrand's Science Series, No. 56.

**Boussinesq.** Complément à de précédentes notes sur la poussée des terres. *Annales P. et C., 1884.

**Bousin.** Equilibrium of pulverulent bodies. The **equilibrium of earth** when confined by a wall. †Van N., 1881.

**Cain.** Modification of Weyrauch's Theory. Van N., 1880.

—— **Earth-pressure.** Modification of Weyrauch's Theory. **Criticism of Baker's articles.** Van N., 1882.

—— Uniform cross-section, and T abutments: their proper proportions and sizes, deduced from Rankine's general formulas. Van N., 1872.

—— **Practical designing of** retaining-walls. Van N. Science Series, No. 3, 1888.

**Chaperon.** Observations sur le mémoire de M. de Sazilly (1851). Stabilité et consolidation des talus. Annales P. et C., 1853.

**Considère.** Note sur la poussée des terres. Annales P. et C., 1870.

**Cousinery.** Détermination graphique de l'épaisseur des murs de soutènement. Annales P. et C., 1841.

**De Lafont.** Sur la poussée des terres et sur les dimensions à donner, suivant leurs profils, aux murs de soutènement et de reservoirs d'eau. Annales P. et C., 1866.

**De Sazilly.** Sur les conditions d'équilibre des massifs de terre, et sur les revêtements des talus. Annales P. et C., 1851.

**Eddy.** Retaining-walls treated graphically. Van N., 1877.

**Flamant.** Note sur la poussée des terres. Annales P. et C., 1882.

—— Résumé d'articles publiés par la Société des Ingénieures Civils de Londres sur la poussée des terres. Annales P. et C., 1883.

---

\* **Annales des** Ponts et Chaussées.
† **Van** Nostrand's Magazine.

FLAMANT. Note sur la poussée des terres. Annales P. et C., 1872.

—— Mémoire sur la stabilité de la terre sans cohésion par W. J. Macquorm Rankine (Extrait 1856-57). Annales P. et C., 1874.

GOBIN. Détermination précis de la stabilité des murs de soutènement et de la poussée des terres. Annales P. et C., 1883.

GOULD. Theory of J. Dubosque. Van N., 1883.

—— Designing. Van N., 1877.

JACOB. Practical designing of retaining-walls. Van N., 1873; also Van N. Science Series, No. 3.

JACQUIER. Note sur la détermination graphique de la poussée des terres. Annales P. et C., 1882.

KLEITZ. Détermination de la poussée des terres et établissement des murs de soutènement. Annales P. et C., 1844.

LAGREUE. Note sur la poussée des terres avec ou sans surcharges. Annales P. et C., 1881.

L'ÈVEILLÉ. De l'emploi des contre-forts. Annales P. et C. 1844.

LEYGUE. Sur les grands murs de soutènement de la ligne de Mezamet a Bédarieux. Annales P. et C., 1887.

—— Nouvelle recherche sur la poussée des terres et le profil de revêtement le plus économique. Annales P. et C., 1885.

MERRIMAN. On the theories of the lateral pressure of sand against retaining walls. (School of Mines Quarterly.) Engineering News, 1888.

—— The theory and calculation of earthwork. Engineering News, 1885.

REBHANN. Theorie des Erddruckes und der Futtermauern. Wien, 1870 and 1871.

SAINT-GUILHEM. Sur la poussée des terres avec ou sans surcharge. Annales P. et C., 1858.

SCHEFFLER-FOURNIE. Traité de la stabilité des constructions. Paris, 1864.

TATE. Surcharged and different forms of retaining walls. Van N., 1873; also Van N. Science Series, No. Also published by E. & F. N. Spon.

THORNTON. Theory. Van N., 1879.

## FOUNDATIONS.

BAKER. A treatise on masonry construction. John Wiley & Sons, N. Y.

PATTON. A practical treatise on foundations. John Wiley & Sons, N. Y.

—— A treatise on civil engineering. John Wiley & Sons, N. Y.

For articles in engineering periodicals the reader is referred to a "Descriptive Index of Current Engineering Literature" (1884–1891), published by the Board of Managers of the Association of Engineering Societies.

## DIAGRAM I.

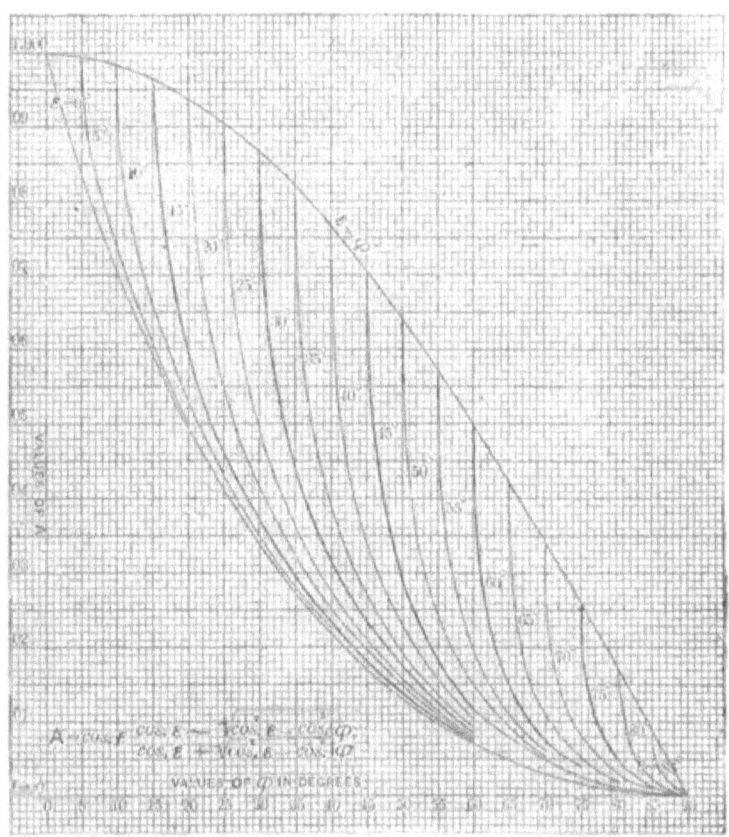

# TABLES.

*Table I* contains the crushing-strengths and the average weights of stone likely to be used in the construction of retaining-walls and foundations; also the average weights of different earths.

*Table II* contains the coefficients of friction, limiting angles of friction, and the reciprocals of the coefficients of friction for various substances.

*Tables III, IV, and V* contain the values of the coefficients [see equation (1′)] $(B)$, $(C)$, $(D)$ and $(E)$, where

$$(B) = \frac{\cos(\epsilon - \alpha)}{\cos^2\alpha \cos \epsilon}, \quad (C) = \sin^2 \alpha, \quad (D) = \left\{ \frac{\cos(\epsilon - \alpha)}{\cos \epsilon} \right\}^2$$

and

$$(E) = 2 \sin \alpha \sin \epsilon \frac{\cos(\epsilon - \alpha)}{\cos \epsilon}.$$

The tables were computed with a Thacher calculating instrument and checked by means of diagrams. It is believed that they are correct to the second place of decimals; an error in the third place of decimals does not affect the results for practical purposes.

*Table VI* contains the natural sines, cosines and tangents.

## TABLE I.

### VALUES OF $W$.

| Name of Substance. | Crushing Lds. in tons per sq. ft. | Average weight in lbs. per cu. ft. |
|---|---|---|
| Alabaster............ | ............ | 144 |
| Brick, best pressed............ | 40 to 300 | 150 |
| " common hard............ | ............ | 125 |
| " soft inferior............ | ............ | 100 |
| Chalk............ | 20 to 30 | 156 |
| Cement, loose............ | ............ | 49.6 to 102 |
| Flint............ | ............ | 162 |
| Feldspar............ | ............ | 166 |
| Granite............ | 300 to 1200 | 170 |
| Gneiss............ | ............ | 168 |
| Greenstone, trap............ | ............ | 187 |
| Hornblende, black............ | ............ | 203 |
| Limestones and Marbles, ordinary...... | 250 to 1000 | { 164.4 <br> { 168 |
| Mortar, hardened............ | ............ | 103 |
| Quartz, common............ | ............ | 165 |
| Sandstone............ | 150 to 550 | 151 |
| Shales............ | ............ | 162 |
| Slate............ | 400 to 800 | 175 |
| Soapstone............ | ............ | 170 |

### VALUES OF $\gamma$.

| Name of Substance. | Average weight in lbs. per cu. ft. |
|---|---|
| Earth, common loam, loose............ | 72 to 80 |
| " " " shaken............ | 82 " 92 |
| " " " rammed moderately............ | 90 " 100 |
| Gravel............ | 90 " 106 |
| Sand............ | 90 " 106 |
| Soft flowing mud............ | 104 " 120 |
| Sand perfectly wet............ | 118 " 129 |

## TABLE II.

### * ANGLES AND COEFFICIENTS OF FRICTION.

|  | tan $\phi$ | $\phi$ | $\dfrac{1}{\tan \phi}$ |
|---|---|---|---|
| Dry masonry and brickwork | 0.6 to 0.7 | 31° to 35° | 1.67 to 1.43 |
| Masonry and brickwork with damp mortar | 0.74 | 36½° | 1.35 |
| Timber on stone | about 0.4 | 22° | 2.5 |
| Iron on stone | 0.7 to 0.3 | 35° to 16⅔° | 1.43 to 3.33 |
| Timber on timber | 0.5 " 0.2 | 26½° " 11¼° | 2 " 5 |
| Timber on metals | 0.6 " 0.2 | 31° " 11½° | 1.67 " 5 |
| Metals on metals | 0.25 " 0.15 | 14° " 8½° | 4 " 6.67 |
| Masonry on dry clay | 0.51 | 27° | 1.96 |
| "    " moist clay | 0.33 | 18¼° | 3. |
| Earth on earth | 0.25 to 1.0 | 14° to 45° | 4 to 1 |
| Earth on earth, dry sand, clay, and mixed earth | 0.38 " 0.75 | 21° " 37° | 2.63 " 1.33 |
| Earth on earth, damp clay | 1.0 | 45° | 1 |
| Earth on earth, wet clay | 0.31 | 17° | 3.23 |
| Earth on earth, shingle and gravel | 0.81 | 39° to 48° | 1.23 to 0.9 |

* From Rankine's Applied Mechanics.

## TABLE III.

| ε | $a = 5°$ | $a = 6°$ | $a = 7°$ | $a = 8°$ | $a = 9°$ |
|---|---|---|---|---|---|
|   | (B) | (B) | (B) | (B) | (B) |
| 0 | 1.004 | 1.005 | 1.007 | 1.010 | 1.012 |
| 5 | 1.012 | 1.015 | 1.018 | 1.022 | 1.026 |
| 10 | 1.019 | 1.024 | 1.029 | 1.035 | 1.040 |
| 15 | 1.027 | 1.034 | 1.041 | 1.048 | 1.055 |
| 20 | 1.036 | 1.044 | 1.052 | 1.062 | 1.071 |
| 25 | 1.045 | 1.055 | 1.065 | 1.076 | 1.088 |
| 30 | 1.055 | 1.066 | 1.079 | 1.092 | 1.105 |
| 35 | 1.065 | 1.079 | 1.094 | 1.109 | 1.124 |
| 40 | 1.078 | 1.094 | 1.111 | 1.129 | 1.147 |
| 45 | 1.093 | 1.111 | 1.131 | 1.152 | 1.173 |
|   | (C) | (C) | (C) | (C) | (C) |
|   | 0.008 | 0.011 | 0.015 | 0.019 | 0.024 |

## TABLE IV.

| ε | $a = 5°$ | $a = 6°$ | $a = 7°$ | $a = 8°$ | $a = 9°$ |
|---|---|---|---|---|---|
|   | (D) | (D) | (D) | (D) | (D) |
| 0 | 0.992 | 0.989 | 0.985 | 0.981 | 0.976 |
| 5 | 1.008 | 1.008 | 1.006 | 1.005 | 1.003 |
| 10 | 1.023 | 1.026 | 1.028 | 1.030 | 1.031 |
| 15 | 1.040 | 1.046 | 1.051 | 1.056 | 1.060 |
| 20 | 1.057 | 1.066 | 1.075 | 1.084 | 1.092 |
| 25 | 1.075 | 1.089 | 1.102 | 1.114 | 1.125 |
| 30 | 1.096 | 1.113 | 1.130 | 1.147 | 1.163 |
| 35 | 1.118 | 1.140 | 1.164 | 1.183 | 1.204 |
| 40 | 1.144 | 1.172 | 1.199 | 1.226 | 1.253 |
| 45 | 1.174 | 1.208 | 1.242 | 1.276 | 1.309 |

## TABLE V.

| ε | $a = 5°$ | $a = 6°$ | $a = 7°$ | $a = 8°$ | $a = 9°$ |
|---|---|---|---|---|---|
|   | (E) | (E) | (E) | (E) | (E) |
| 0 | 0 | 0 | 0 | 0 | 0 |
| 5 | 0.015 | 0.018 | 0.021 | 0.024 | 0.027 |
| 10 | 0.031 | 0.037 | 0.043 | 0.049 | 0.055 |
| 15 | 0.046 | 0.055 | 0.065 | 0.074 | 0.083 |
| 20 | 0.061 | 0.074 | 0.086 | 0.099 | 0.112 |
| 25 | 0.076 | 0.092 | 0.108 | 0.124 | 0.140 |
| 30 | 0.091 | 0.110 | 0.130 | 0.149 | 0.169 |
| 35 | 0.106 | 0.128 | 0.151 | 0.174 | 0.197 |
| 40 | 0.120 | 0.145 | 0.172 | 0.198 | 0.225 |
| 45 | 0.134 | 0.162 | 0.192 | 0.222 | 0.253 |

## TABLE III—*Continued.*

| ϵ | $a = 10°$ (B) | $a = 11°$ (B) | $a = 12°$ (B) | $a = 13°$ (B) | $a = 14°$ (B) |
|---|---|---|---|---|---|
| 0  | 1.015 | 1.019 | 1.022 | 1.026 | 1.031 |
| 5  | 1.031 | 1.037 | 1.041 | 1.047 | 1.053 |
| 10 | 1.046 | 1.055 | 1.061 | 1.068 | 1.076 |
| 15 | 1.063 | 1.073 | 1.081 | 1.090 | 1.100 |
| 20 | 1.081 | 1.092 | 1.103 | 1.112 | 1.125 |
| 25 | 1.099 | 1.112 | 1.124 | 1.136 | 1.150 |
| 30 | 1.119 | 1.135 | 1.151 | 1.163 | 1.179 |
| 35 | 1.141 | 1.159 | 1.175 | 1.195 | 1.211 |
| 40 | 1.166 | 1.186 | 1.205 | 1.225 | 1.245 |
| 45 | 1.195 | 1.218 | 1.240 | 1.263 | 1.288 |
| (C) | 0.030 | 0.036 | 0.043 | 0.051 | 0.029 |

## TABLE IV—*Continued.*

| ϵ | $a = 10°$ (D) | $a = 11°$ (D) | $a = 12°$ (D) | $a = 13°$ (D) | $a = 14°$ (D) |
|---|---|---|---|---|---|
| 0  | 0.970 | 0.964 | 0.957 | 0.950 | 0.943 |
| 5  | 1.000 | 0.997 | 0.993 | 0.988 | 0.983 |
| 10 | 1.031 | 1.031 | 1.030 | 1.028 | 1.026 |
| 15 | 1.064 | 1.067 | 1.069 | 1.061 | 1.072 |
| 20 | 1.099 | 1.105 | 1.110 | 1.116 | 1.121 |
| 25 | 1.136 | 1.147 | 1.156 | 1.165 | 1.173 |
| 30 | 1.178 | 1.194 | 1.204 | 1.220 | 1.232 |
| 35 | 1.224 | 1.244 | 1.262 | 1.281 | 1.300 |
| 40 | 1.291 | 1.304 | 1.328 | 1.353 | 1.377 |
| 45 | 1.342 | 1.375 | 1.407 | 1.438 | 1.469 |

## TABLE V—*Continued.*

| ϵ | $a = 10°$ (E) | $a = 11°$ (E) | $a = 12°$ (E) | $a = 13°$ (E) | $a = 14°$ (E) |
|---|---|---|---|---|---|
| 0  | 0 | 0 | 0 | 0 | 0 |
| 5  | 0.030 | 0.032 | 0.036 | 0.039 | 0.042 |
| 10 | 0.061 | 0.067 | 0.073 | 0.079 | 0.085 |
| 15 | 0.093 | 0.102 | 0.111 | 0.119 | 0.130 |
| 20 | 0.124 | 0.137 | 0.150 | 0.163 | 0.175 |
| 25 | 0.156 | 0.173 | 0.189 | 0.205 | 0.221 |
| 30 | 0.188 | 0.208 | 0.216 | 0.248 | 0.269 |
| 35 | 0.220 | 0.244 | 0.268 | 0.292 | 0.316 |
| 40 | 0.252 | 0.280 | 0.308 | 0.336 | 0.365 |
| 45 | 0.284 | 0.316 | 0.349 | 0.382 | 0.415 |

## TABLE III—*Continued.*

| ε | a = 15° (B) | a = 16° (B) | a = 17° (B) | a = 18° (B) | a = 20° (B) |
|---|---|---|---|---|---|
| 0 | 1.035 | 1.040 | 1.048 | 1.051 | 1.062 |
| 5 | 1.059 | 1.066 | 1.076 | 1.081 | 1.098 |
| 10 | 1.084 | 1.093 | 1.104 | 1.112 | 1.132 |
| 15 | 1.110 | 1.120 | 1.134 | 1.138 | 1.168 |
| 20 | 1.135 | 1.149 | 1.165 | 1.177 | 1.218 |
| 25 | 1.165 | 1.179 | 1.197 | 1.211 | 1.245 |
| 30 | 1.195 | 1.212 | 1.233 | 1.248 | 1.288 |
| 35 | 1.229 | 1.249 | 1.272 | 1.291 | 1.339 |
| 40 | 1.268 | 1.291 | 1.317 | 1.340 | 1.389 |
| 45 | 1.313 | 1.338 | 1.369 | 1.393 | 1.451 |
| | (C) | (C) | (C) | (C) | (C) |
| | 0.067 | 0.076 | 0.086 | 0.095 | 0.117 |

## TABLE IV—*Continued.*

| ε | a = 15° (D) | a = 16° (D) | a = 17° (D) | a = 18° (D) | a = 20° (D) |
|---|---|---|---|---|---|
| 0 | 0.933 | 0.924 | 0.915 | 0.905 | 0.883 |
| 5 | 0.977 | 0.971 | 0.964 | 0.957 | 0.940 |
| 10 | 1.023 | 1.018 | 1.016 | 1.011 | 1.000 |
| 15 | 1.072 | 1.073 | 1.071 | 1.069 | 1.068 |
| 20 | 1.124 | 1.127 | 1.129 | 1.131 | 1.132 |
| 25 | 1.181 | 1.188 | 1.194 | 1.200 | 1.208 |
| 30 | 1.244 | 1.256 | 1.266 | 1.276 | 1.293 |
| 35 | 1.316 | 1.332 | 1.348 | 1.363 | 1.390 |
| 40 | 1.400 | 1.422 | 1.444 | 1.465 | 1.505 |
| 45 | 1.500 | 1.530 | 1.559 | 1.588 | 1.643 |

## TABLE V—*Continued.*

| ε | a = 15° (E) | a = 16° (E) | a = 17° (E) | a = 18° (E) | a = 20° (E) |
|---|---|---|---|---|---|
| 0 | 0 | 0 | 0 | 0 | 0 |
| 5 | 0.045 | 0.047 | 0.050 | 0.053 | 0.058 |
| 10 | 0.091 | 0.097 | 0.102 | 0.108 | 0.119 |
| 15 | 0.139 | 0.148 | 0.157 | 0.165 | 0.183 |
| 20 | 0.188 | 0.200 | 0.213 | 0.225 | 0.249 |
| 25 | 0.238 | 0.254 | 0.270 | 0.177 | 0.318 |
| 30 | 0.289 | 0.309 | 0.329 | 0.349 | 0.389 |
| 35 | 0.341 | 0.365 | 0.390 | 0.414 | 0.463 |
| 40 | 0.394 | 0.423 | 0.452 | 0.481 | 0.539 |
| 45 | 0.448 | 0.482 | 0.516 | 0.551 | 0.620 |

# TABLE VI.

NATURAL SINES, COSINES, TANGENTS AND COTANGENTS.

## NATURAL SINES AND COSINES.

| ′ | 0° | | 1° | | 2° | | 3° | | 4° | | ′ |
|---|---|---|---|---|---|---|---|---|---|---|---|
|   | Sine | Cosin | Sine | Cosin | Sine | Cosin | Sine | Cosin | Sine | Cosin |   |
| 0 | .00000 | One. | .01745 | .99985 | .03490 | .99939 | .05234 | .99863 | .06976 | .99756 | 60 |
| 1 | .00029 | One. | .01774 | .99984 | .03519 | .99938 | .05263 | .99861 | .07005 | .99754 | 59 |
| 2 | .00058 | One. | .01803 | .99984 | .03548 | .99937 | .05292 | .99860 | .07034 | .99752 | 58 |
| 3 | .00087 | One. | .01832 | .99983 | .03577 | .99936 | .05321 | .99858 | .07063 | .99750 | 57 |
| 4 | .00116 | One. | .01862 | .99983 | .03606 | .99935 | .05350 | .99857 | .07092 | .99748 | 56 |
| 5 | .00145 | One. | .01891 | .99982 | .03635 | .99934 | .05379 | .99855 | .07121 | .99746 | 55 |
| 6 | .00175 | One. | .01920 | .99982 | .03664 | .99933 | .05408 | .99854 | .07150 | .99744 | 54 |
| 7 | .00204 | One. | .01949 | .99981 | .03693 | .99932 | .05437 | .99852 | .07179 | .99742 | 53 |
| 8 | .00233 | One. | .01978 | .99981 | .03723 | .99931 | .05466 | .99851 | .07208 | .99740 | 52 |
| 9 | .00262 | One. | .02007 | .99980 | .03752 | .99930 | .05495 | .99849 | .07237 | .99738 | 51 |
| 10 | .00291 | One. | .02036 | .99979 | .03781 | .99929 | .05524 | .99847 | .07266 | .99736 | 50 |
| 11 | .00320 | .99999 | .02065 | .99979 | .03810 | .99927 | .05553 | .99846 | .07295 | .99734 | 49 |
| 12 | .00349 | .99999 | .02094 | .99978 | .03839 | .99926 | .05582 | .99844 | .07324 | .99731 | 48 |
| 13 | .00378 | .99999 | .02123 | .99977 | .03868 | .99925 | .05611 | .99842 | .07353 | .99729 | 47 |
| 14 | .00407 | .99999 | .02152 | .99977 | .03897 | .99924 | .05640 | .99841 | .07382 | .99727 | 46 |
| 15 | .00436 | .99999 | .02181 | .99976 | .03926 | .99923 | .05669 | .99839 | .07411 | .99725 | 45 |
| 16 | .00465 | .99999 | .02211 | .99976 | .03955 | .99922 | .05698 | .99838 | .07440 | .99723 | 44 |
| 17 | .00495 | .99999 | .02240 | .99975 | .03984 | .99921 | .05727 | .99836 | .07469 | .99721 | 43 |
| 18 | .00524 | .99999 | .02269 | .99974 | .04013 | .99919 | .05756 | .99834 | .07498 | .99719 | 42 |
| 19 | .00553 | .99998 | .02298 | .99974 | .04042 | .99918 | .05785 | .99833 | .07527 | .99716 | 41 |
| 20 | .00582 | .99998 | .02327 | .99973 | .04071 | .99917 | .05814 | .99831 | .07556 | .99714 | 40 |
| 21 | .00611 | .99998 | .02356 | .99972 | .04100 | .99916 | .05844 | .99829 | .07585 | .99712 | 39 |
| 22 | .00640 | .99998 | .02385 | .99972 | .04129 | .99915 | .05873 | .99827 | .07614 | .99710 | 38 |
| 23 | .00669 | .99998 | .02414 | .99971 | .04159 | .99913 | .05902 | .99826 | .07643 | .99708 | 37 |
| 24 | .00698 | .99998 | .02443 | .99970 | .04188 | .99912 | .05931 | .99824 | .07672 | .99705 | 36 |
| 25 | .00727 | .99997 | .02472 | .99969 | .04217 | .99911 | .05960 | .99822 | .07701 | .99703 | 35 |
| 26 | .00756 | .99997 | .02501 | .99969 | .04246 | .99910 | .05989 | .99821 | .07730 | .99701 | 34 |
| 27 | .00785 | .99997 | .02530 | .99968 | .04275 | .99909 | .06018 | .99819 | .07759 | .99699 | 33 |
| 28 | .00814 | .99997 | .02560 | .99967 | .04304 | .99907 | .06047 | .99817 | .07788 | .99696 | 32 |
| 29 | .00844 | .99996 | .02589 | .99966 | .04333 | .99906 | .06076 | .99815 | .07817 | .99694 | 31 |
| 30 | .00873 | .99996 | .02618 | .99966 | .04362 | .99905 | .06105 | .99813 | .07846 | .99692 | 30 |
| 31 | .00902 | .99996 | .02647 | .99965 | .04391 | .99904 | .06134 | .99812 | .07875 | .99689 | 29 |
| 32 | .00931 | .99996 | .02676 | .99964 | .04420 | .99902 | .06163 | .99810 | .07904 | .99687 | 28 |
| 33 | .00960 | .99995 | .02705 | .99963 | .04449 | .99901 | .06192 | .99808 | .07933 | .99685 | 27 |
| 34 | .00989 | .99995 | .02734 | .99963 | .04478 | .99900 | .06221 | .99806 | .07962 | .99683 | 26 |
| 35 | .01018 | .99995 | .02763 | .99962 | .04507 | .99898 | .06250 | .99804 | .07991 | .99680 | 25 |
| 36 | .01047 | .99995 | .02792 | .99961 | .04536 | .99897 | .06279 | .99803 | .08020 | .99678 | 24 |
| 37 | .01076 | .99994 | .02821 | .99960 | .04565 | .99896 | .06308 | .99801 | .08049 | .99676 | 23 |
| 38 | .01105 | .99994 | .02850 | .99959 | .04594 | .99894 | .06337 | .99799 | .08078 | .99673 | 22 |
| 39 | .01134 | .99994 | .02879 | .99959 | .04623 | .99893 | .06366 | .99797 | .08107 | .99671 | 21 |
| 40 | .01164 | .99993 | .02908 | .99958 | .04653 | .99892 | .06395 | .99795 | .08136 | .99668 | 20 |
| 41 | .01193 | .99993 | .02938 | .99957 | .04682 | .99890 | .06424 | .99793 | .08165 | .99666 | 19 |
| 42 | .01222 | .99993 | .02967 | .99956 | .04711 | .99889 | .06453 | .99792 | .08194 | .99664 | 18 |
| 43 | .01251 | .99992 | .02996 | .99955 | .04740 | .99888 | .06482 | .99790 | .08223 | .99661 | 17 |
| 44 | .01280 | .99992 | .03025 | .99954 | .04769 | .99886 | .06511 | .99788 | .08252 | .99659 | 16 |
| 45 | .01309 | .99991 | .03054 | .99953 | .04798 | .99885 | .06540 | .99786 | .08281 | .99657 | 15 |
| 46 | .01338 | .99991 | .03083 | .99952 | .04827 | .99883 | .06569 | .99784 | .08310 | .99654 | 14 |
| 47 | .01367 | .99991 | .03112 | .99952 | .04856 | .99882 | .06598 | .99782 | .08339 | .99652 | 13 |
| 48 | .01396 | .99990 | .03141 | .99951 | .04885 | .99881 | .06627 | .99780 | .08368 | .99649 | 12 |
| 49 | .01425 | .99990 | .03170 | .99950 | .04914 | .99879 | .06656 | .99778 | .08397 | .99647 | 11 |
| 50 | .01454 | .99989 | .03199 | .99949 | .04943 | .99878 | .06685 | .99776 | .08426 | .99644 | 10 |
| 51 | .01483 | .99989 | .03228 | .99948 | .04972 | .99876 | .06714 | .99774 | .08455 | .99642 | 9 |
| 52 | .01513 | .99989 | .03257 | .99947 | .05001 | .99875 | .06743 | .99772 | .08484 | .99639 | 8 |
| 53 | .01542 | .99988 | .03286 | .99946 | .05030 | .99873 | .06773 | .99770 | .08513 | .99637 | 7 |
| 54 | .01571 | .99988 | .03316 | .99945 | .05059 | .99872 | .06802 | .99768 | .08542 | .99635 | 6 |
| 55 | .01600 | .99987 | .03345 | .99944 | .05088 | .99870 | .06831 | .99766 | .08571 | .99632 | 5 |
| 56 | .01629 | .99987 | .03374 | .99943 | .05117 | .99869 | .06860 | .99764 | .08600 | .99630 | 4 |
| 57 | .01658 | .99986 | .03403 | .99942 | .05146 | .99867 | .06889 | .99762 | .08629 | .99627 | 3 |
| 58 | .01687 | .99986 | .03432 | .99941 | .05175 | .99866 | .06918 | .99760 | .08658 | .99625 | 2 |
| 59 | .01716 | .99985 | .03461 | .99940 | .05205 | .99864 | .06947 | .99758 | .08687 | .99622 | 1 |
| 60 | .01745 | .99985 | .03490 | .99939 | .05234 | .99863 | .06976 | .99756 | .08716 | .99619 | 0 |
|   | Cosin | Sine | Cosin | Sine | Cosin | Sine | Cosin | Sine | Cosin | Sine |   |
|   | 89° | | 88° | | 87° | | 86° | | 85° | | ′ |

## NATURAL SINES AND COSINES. 113

| ′ | 5° Sine | 5° Cosin | 6° Sine | 6° Cosin | 7° Sine | 7° Cosin | 8° Sine | 8° Cosin | 9° Sine | 9° Cosin | ′ |
|---|---|---|---|---|---|---|---|---|---|---|---|
| 0 | .08716 | .99619 | .10453 | .99452 | .12187 | .99255 | .13917 | .99027 | .15643 | .98769 | 60 |
| 1 | .08745 | .99617 | .10482 | .99449 | .12216 | .99251 | .13946 | .99023 | .15672 | .98764 | 59 |
| 2 | .08774 | .99614 | .10511 | .99446 | .12245 | .99248 | .13975 | .99019 | .15701 | .98760 | 58 |
| 3 | .08803 | .99612 | .10540 | .99443 | .12274 | .99244 | .14004 | .99015 | .15730 | .98755 | 57 |
| 4 | .08831 | .99609 | .10569 | .99440 | .12302 | .99240 | .14033 | .99011 | .15758 | .98751 | 56 |
| 5 | .08860 | .99607 | .10597 | .99437 | .12331 | .99237 | .14061 | .99006 | .15787 | .98746 | 55 |
| 6 | .08889 | .99604 | .10626 | .99434 | .12360 | .99233 | .14090 | .99002 | .15816 | .98741 | 54 |
| 7 | .08918 | .99602 | .10655 | .99431 | .12389 | .99230 | .14119 | .98998 | .15845 | .98737 | 53 |
| 8 | .08947 | .99599 | .10684 | .99428 | .12418 | .99226 | .14148 | .98994 | .15873 | .98732 | 52 |
| 9 | .08976 | .99596 | .10713 | .99424 | .12447 | .99222 | .14177 | .98990 | .15902 | .98728 | 51 |
| 10 | .09005 | .99594 | .10742 | .99421 | .12476 | .99219 | .14205 | .98986 | .15931 | .98723 | 50 |
| 11 | .09034 | .99591 | .10771 | .99418 | .12504 | .99215 | .14234 | .98982 | .15959 | .98718 | 49 |
| 12 | .09063 | .99588 | .10800 | .99415 | .12533 | .99211 | .14263 | .98978 | .15988 | .98714 | 48 |
| 13 | .09092 | .99586 | .10829 | .99412 | .12562 | .99208 | .14292 | .98973 | .16017 | .98709 | 47 |
| 14 | .09121 | .99583 | .10858 | .99409 | .12591 | .99204 | .14320 | .98969 | .16046 | .98704 | 46 |
| 15 | .09150 | .99580 | .10887 | .99406 | .12620 | .99200 | .14349 | .98965 | .16074 | .98700 | 45 |
| 16 | .09179 | .99578 | .10916 | .99402 | .12649 | .99197 | .14378 | .98961 | .16103 | .98695 | 44 |
| 17 | .09208 | .99575 | .10945 | .99399 | .12678 | .99193 | .14407 | .98957 | .16132 | .98690 | 43 |
| 18 | .09237 | .99572 | .10973 | .99396 | .12706 | .99189 | .14436 | .98953 | .16160 | .98686 | 42 |
| 19 | .09266 | .99570 | .11002 | .99393 | .12735 | .99186 | .14464 | .98948 | .16189 | .98681 | 41 |
| 20 | .09295 | .99567 | .11031 | .99390 | .12764 | .99182 | .14493 | .98944 | .16218 | .98676 | 40 |
| 21 | .09324 | .99564 | .11060 | .99386 | .12793 | .99178 | .14522 | .98940 | .16246 | .98671 | 39 |
| 22 | .09353 | .99562 | .11089 | .99383 | .12822 | .99175 | .14551 | .98936 | .16275 | .98667 | 38 |
| 23 | .09382 | .99559 | .11118 | .99380 | .12851 | .99171 | .14580 | .98931 | .16304 | .98662 | 37 |
| 24 | .09411 | .99556 | .11147 | .99377 | .12880 | .99167 | .14608 | .98927 | .16333 | .98657 | 36 |
| 25 | .09440 | .99553 | .11176 | .99374 | .12908 | .99163 | .14637 | .98923 | .16361 | .98652 | 35 |
| 26 | .09469 | .99551 | .11205 | .99370 | .12937 | .99160 | .14666 | .98919 | .16390 | .98648 | 34 |
| 27 | .09498 | .99548 | .11234 | .99367 | .12966 | .99156 | .14695 | .98914 | .16419 | .98643 | 33 |
| 28 | .09527 | .99545 | .11263 | .99364 | .12995 | .99152 | .14723 | .98910 | .16447 | .98638 | 32 |
| 29 | .09556 | .99542 | .11291 | .99360 | .13024 | .99148 | .14752 | .98906 | .16476 | .98633 | 31 |
| 30 | .09585 | .99540 | .11320 | .99357 | .13053 | .99144 | .14781 | .98902 | .16505 | .98629 | 30 |
| 31 | .09614 | .99537 | .11349 | .99354 | .13081 | .99141 | .14810 | .98897 | .16533 | .98624 | 29 |
| 32 | .09642 | .99534 | .11378 | .99351 | .13110 | .99137 | .14838 | .98893 | .16562 | .98619 | 28 |
| 33 | .09671 | .99531 | .11407 | .99347 | .13139 | .99133 | .14867 | .98889 | .16591 | .98614 | 27 |
| 34 | .09700 | .99528 | .11436 | .99344 | .13168 | .99129 | .14896 | .98884 | .16620 | .98609 | 26 |
| 35 | .09729 | .99526 | .11465 | .99341 | .13197 | .99125 | .14925 | .98880 | .16648 | .98604 | 25 |
| 36 | .09758 | .99523 | .11494 | .99337 | .13226 | .99122 | .14954 | .98876 | .16677 | .98600 | 24 |
| 37 | .09787 | .99520 | .11523 | .99334 | .13254 | .99118 | .14982 | .98871 | .16706 | .98595 | 23 |
| 38 | .09816 | .99517 | .11552 | .99331 | .13283 | .99114 | .15011 | .98867 | .16734 | .98590 | 22 |
| 39 | .09845 | .99514 | .11580 | .99327 | .13312 | .99110 | .15040 | .98863 | .16763 | .98585 | 21 |
| 40 | .09874 | .99511 | .11609 | .99324 | .13341 | .99106 | .15069 | .98858 | .16792 | .98580 | 20 |
| 41 | .09903 | .99508 | .11638 | .99320 | .13370 | .99102 | .15097 | .98854 | .16820 | .98575 | 19 |
| 42 | .09932 | .99506 | .11667 | .99317 | .13399 | .99098 | .15126 | .98849 | .16849 | .98570 | 18 |
| 43 | .09961 | .99503 | .11696 | .99314 | .13427 | .99094 | .15155 | .98845 | .16878 | .98565 | 17 |
| 44 | .09990 | .99500 | .11725 | .99310 | .13456 | .99091 | .15184 | .98841 | .16906 | .98561 | 16 |
| 45 | .10019 | .99497 | .11754 | .99307 | .13485 | .99087 | .15212 | .98836 | .16935 | .98556 | 15 |
| 46 | .10048 | .99494 | .11783 | .99303 | .13514 | .99083 | .15241 | .98832 | .16964 | .98551 | 14 |
| 47 | .10077 | .99491 | .11812 | .99300 | .13543 | .99079 | .15270 | .98827 | .16992 | .98546 | 13 |
| 48 | .10106 | .99488 | .11840 | .99297 | .13572 | .99075 | .15299 | .98823 | .17021 | .98541 | 12 |
| 49 | .10135 | .99485 | .11869 | .99293 | .13600 | .99071 | .15327 | .98818 | .17050 | .98536 | 11 |
| 50 | .10164 | .99482 | .11898 | .99290 | .13629 | .99067 | .15356 | .98814 | .17078 | .98531 | 10 |
| 51 | .10192 | .99479 | .11927 | .99286 | .13658 | .99063 | .15385 | .98809 | .17107 | .98526 | 9 |
| 52 | .10221 | .99476 | .11956 | .99283 | .13687 | .99059 | .15414 | .98805 | .17136 | .98521 | 8 |
| 53 | .10250 | .99473 | .11985 | .99279 | .13716 | .99055 | .15442 | .98800 | .17164 | .98516 | 7 |
| 54 | .10279 | .99470 | .12014 | .99276 | .13744 | .99051 | .15471 | .98796 | .17193 | .98511 | 6 |
| 55 | .10308 | .99467 | .12043 | .99272 | .13773 | .99047 | .15500 | .98791 | .17222 | .98506 | 5 |
| 56 | .10337 | .99464 | .12071 | .99269 | .13802 | .99043 | .15529 | .98787 | .17250 | .98501 | 4 |
| 57 | .10366 | .99461 | .12100 | .99265 | .13831 | .99039 | .15557 | .98782 | .17279 | .98496 | 3 |
| 58 | .10395 | .99458 | .12129 | .99262 | .13860 | .99035 | .15586 | .98778 | .17308 | .98491 | 2 |
| 59 | .10424 | .99455 | .12158 | .99258 | .13889 | .99031 | .15615 | .98773 | .17336 | .98486 | 1 |
| 60 | .10453 | .99452 | .12187 | .99255 | .13917 | .99027 | .15643 | .98769 | .17365 | .98481 | 0 |
| ′ | Cosin | Sine | Cosin | Sine | Cosin | Sine | Cosin | Sine | Cosin | Sine | ′ |
|  | 84° | | 83° | | 82° | | 81° | | 80° | |  |

## NATURAL SINES AND COSINES.

| ′ | 10° Sine | Cosin | 11° Sine | Cosin | 12° Sine | Cosin | 13° Sine | Cosin | 14° Sine | Cosin | ′ |
|---|---|---|---|---|---|---|---|---|---|---|---|
| 0 | .17365 | .98481 | .19081 | .98163 | .20791 | .97815 | .22495 | .97437 | .24192 | .97030 | 60 |
| 1 | .17393 | .98476 | .19109 | .98157 | .20820 | .97809 | .22523 | .97430 | .24220 | .97023 | 59 |
| 2 | .17422 | .98471 | .19138 | .98152 | .20848 | .97803 | .22552 | .97424 | .24249 | .97015 | 58 |
| 3 | .17451 | .98466 | .19167 | .98146 | .20877 | .97797 | .22580 | .97417 | .24277 | .97008 | 57 |
| 4 | .17479 | .98461 | .19195 | .98140 | .20905 | .97791 | .22608 | .97411 | .24305 | .97001 | 56 |
| 5 | .17508 | .98455 | .19224 | .98135 | .20933 | .97784 | .22637 | .97404 | .24333 | .96994 | 55 |
| 6 | .17537 | .98450 | .19252 | .98129 | .20962 | .97778 | .22665 | .97398 | .24362 | .96987 | 54 |
| 7 | .17565 | .98445 | .19281 | .98124 | .20990 | .97772 | .22693 | .97391 | .24390 | .96980 | 53 |
| 8 | .17594 | .98440 | .19309 | .98118 | .21019 | .97766 | .22722 | .97384 | .24418 | .96973 | 52 |
| 9 | .17623 | .98435 | .19338 | .98112 | .21047 | .97760 | .22750 | .97378 | .24446 | .96966 | 51 |
| 10 | .17651 | .98430 | .19366 | .98107 | .21076 | .97754 | .22778 | .97371 | .24474 | .96959 | 50 |
| 11 | .17680 | .98425 | .19395 | .98101 | .21104 | .97748 | .22807 | .97365 | .24503 | .96952 | 49 |
| 12 | .17708 | .98420 | .19423 | .98096 | .21132 | .97742 | .22835 | .97358 | .24531 | .96945 | 48 |
| 13 | .17737 | .98414 | .19452 | .98090 | .21161 | .97735 | .22863 | .97351 | .24559 | .96937 | 47 |
| 14 | .17766 | .98409 | .19481 | .98084 | .21189 | .97729 | .22892 | .97345 | .24587 | .96930 | 46 |
| 15 | .17794 | .98404 | .19509 | .98079 | .21218 | .97723 | .22920 | .97338 | .24615 | .96923 | 45 |
| 16 | .17823 | .98399 | .19538 | .98073 | .21246 | .97717 | .22948 | .97331 | .24644 | .96916 | 44 |
| 17 | .17852 | .98394 | .19566 | .98067 | .21275 | .97711 | .22977 | .97325 | .24672 | .96909 | 43 |
| 18 | .17880 | .98389 | .19595 | .98061 | .21303 | .97705 | .23005 | .97318 | .24700 | .96902 | 42 |
| 19 | .17909 | .98383 | .19623 | .98056 | .21331 | .97698 | .23033 | .97311 | .24728 | .96894 | 41 |
| 20 | .17937 | .98378 | .19652 | .98050 | .21360 | .97692 | .23062 | .97304 | .24756 | .96887 | 40 |
| 21 | .17966 | .98373 | .19680 | .98044 | .21388 | .97686 | .23090 | .97298 | .24784 | .96880 | 39 |
| 22 | .17995 | .98368 | .19709 | .98039 | .21417 | .97680 | .23118 | .97291 | .24813 | .96873 | 38 |
| 23 | .18023 | .98362 | .19737 | .98033 | .21445 | .97673 | .23146 | .97284 | .24841 | .96866 | 37 |
| 24 | .18052 | .98357 | .19766 | .98027 | .21474 | .97667 | .23175 | .97278 | .24869 | .96858 | 36 |
| 25 | .18081 | .98352 | .19794 | .98021 | .21502 | .97661 | .23203 | .97271 | .24897 | .96851 | 35 |
| 26 | .18109 | .98347 | .19823 | .98016 | .21530 | .97655 | .23231 | .97264 | .24925 | .96844 | 34 |
| 27 | .18138 | .98341 | .19851 | .98010 | .21559 | .97648 | .23260 | .97257 | .24954 | .96837 | 33 |
| 28 | .18166 | .98336 | .19880 | .98004 | .21587 | .97642 | .23288 | .97251 | .24982 | .96829 | 32 |
| 29 | .18195 | .98331 | .19908 | .97998 | .21616 | .97636 | .23316 | .97244 | .25010 | .96822 | 31 |
| 30 | .18224 | .98325 | .19937 | .97992 | .21644 | .97630 | .23345 | .97237 | .25038 | .96815 | 30 |
| 31 | .18252 | .98320 | .19965 | .97987 | .21672 | .97623 | .23373 | .97230 | .25066 | .96807 | 29 |
| 32 | .18281 | .98315 | .19994 | .97981 | .21701 | .97617 | .23401 | .97223 | .25094 | .96800 | 28 |
| 33 | .18309 | .98310 | .20022 | .97975 | .21729 | .97611 | .23429 | .97217 | .25122 | .96793 | 27 |
| 34 | .18338 | .98304 | .20051 | .97969 | .21758 | .97604 | .23458 | .97210 | .25151 | .96786 | 26 |
| 35 | .18367 | .98299 | .20079 | .97963 | .21786 | .97598 | .23486 | .97203 | .25179 | .96778 | 25 |
| 36 | .18395 | .98294 | .20108 | .97958 | .21814 | .97592 | .23514 | .97196 | .25207 | .96771 | 24 |
| 37 | .18424 | .98288 | .20136 | .97952 | .21843 | .97585 | .23542 | .97189 | .25235 | .96764 | 23 |
| 38 | .18452 | .98283 | .20165 | .97946 | .21871 | .97579 | .23571 | .97182 | .25263 | .96756 | 22 |
| 39 | .18481 | .98277 | .20193 | .97940 | .21899 | .97573 | .23599 | .97176 | .25291 | .96749 | 21 |
| 40 | .18509 | .98272 | .20222 | .97934 | .21928 | .97566 | .23627 | .97169 | .25320 | .96742 | 20 |
| 41 | .18538 | .98267 | .20250 | .97928 | .21956 | .97560 | .23656 | .97162 | .25348 | .96734 | 19 |
| 42 | .18567 | .98261 | .20279 | .97922 | .21985 | .97553 | .23684 | .97155 | .25376 | .96727 | 18 |
| 43 | .18595 | .98256 | .20307 | .97916 | .22013 | .97547 | .23712 | .97148 | .25404 | .96719 | 17 |
| 44 | .18624 | .98250 | .20336 | .97910 | .22041 | .97541 | .23740 | .97141 | .25432 | .96712 | 16 |
| 45 | .18652 | .98245 | .20364 | .97905 | .22070 | .97534 | .23769 | .97134 | .25460 | .96705 | 15 |
| 46 | .18681 | .98240 | .20393 | .97899 | .22098 | .97528 | .23797 | .97127 | .25488 | .96697 | 14 |
| 47 | .18710 | .98234 | .20421 | .97893 | .22126 | .97521 | .23825 | .97120 | .25516 | .96690 | 13 |
| 48 | .18738 | .98229 | .20450 | .97887 | .22155 | .97515 | .23853 | .97113 | .25545 | .96682 | 12 |
| 49 | .18767 | .98223 | .20478 | .97881 | .22183 | .97508 | .23882 | .97106 | .25573 | .96675 | 11 |
| 50 | .18795 | .98218 | .20507 | .97875 | .22212 | .97502 | .23910 | .97100 | .25601 | .96667 | 10 |
| 51 | .18824 | .98212 | .20535 | .97869 | .22240 | .97496 | .23938 | .97093 | .25629 | .96660 | 9 |
| 52 | .18852 | .98207 | .20563 | .97863 | .22268 | .97489 | .23966 | .97086 | .25657 | .96653 | 8 |
| 53 | .18881 | .98201 | .20592 | .97857 | .22297 | .97483 | .23995 | .97079 | .25685 | .96645 | 7 |
| 54 | .18910 | .98196 | .20620 | .97851 | .22325 | .97476 | .24023 | .97072 | .25713 | .96638 | 6 |
| 55 | .18938 | .98190 | .20649 | .97845 | .22353 | .97470 | .24051 | .97065 | .25741 | .96630 | 5 |
| 56 | .18967 | .98185 | .20677 | .97839 | .22382 | .97463 | .24079 | .97058 | .25769 | .96623 | 4 |
| 57 | .18995 | .98179 | .20706 | .97833 | .22410 | .97457 | .24108 | .97051 | .25798 | .96615 | 3 |
| 58 | .19024 | .98174 | .20734 | .97827 | .22438 | .97450 | .24136 | .97044 | .25826 | .96608 | 2 |
| 59 | .19052 | .98168 | .20763 | .97821 | .22467 | .97444 | .24164 | .97037 | .25854 | .96600 | 1 |
| 60 | .19081 | .98163 | .20791 | .97815 | .22495 | .97437 | .24192 | .97030 | .25882 | .96593 | 0 |
| | Cosin | Sine | Cosin | Sine | Cosin | Sine | Cosin | Sine | Cosin | Sine | ′ |
| | 79° | | 78° | | 77° | | 76° | | 75° | | |

## NATURAL SINES AND COSINES. 115

| ′ | 15° | | 16° | | 17° | | 18° | | 19° | | ′ |
|---|---|---|---|---|---|---|---|---|---|---|---|
| | Sine | Cosin | Sine | Cosin | Sine | Cosin | Sine | Cosin | Sine | Cosin | |
| 0 | .25882 | .96593 | .27564 | .96126 | .29237 | .95630 | .30902 | .95106 | .32557 | .94552 | 60 |
| 1 | .25910 | .96585 | .27592 | .96118 | .29265 | .95623 | .30929 | .95097 | .32584 | .94542 | 59 |
| 2 | .25938 | .96578 | .27620 | .96110 | .29293 | .95613 | .30957 | .95088 | .32612 | .94533 | 58 |
| 3 | .25966 | .96570 | .27648 | .96102 | .29321 | .95605 | .30985 | .95079 | .32639 | .94523 | 57 |
| 4 | .25994 | .96562 | .27676 | .96094 | .29348 | .95596 | .31012 | .95070 | .32667 | .94514 | 56 |
| 5 | .26022 | .96555 | .27704 | .96086 | .29376 | .95588 | .31040 | .95061 | .32694 | .94504 | 55 |
| 6 | .26050 | .96547 | .27731 | .96078 | .29404 | .95579 | .31068 | .95052 | .32722 | .94495 | 54 |
| 7 | .26079 | .96540 | .27759 | .96070 | .29432 | .95571 | .31095 | .95043 | .32749 | .94485 | 53 |
| 8 | .26107 | .96532 | .27787 | .96062 | .29460 | .95562 | .31123 | .95033 | .32777 | .94476 | 52 |
| 9 | .26135 | .96524 | .27815 | .96054 | .29487 | .95554 | .31151 | .95024 | .32804 | .94466 | 51 |
| 10 | .26163 | .96517 | .27843 | .96046 | .29515 | .95545 | .31178 | .95015 | .32832 | .94457 | 50 |
| 11 | .26191 | .96509 | .27871 | .96037 | .29543 | .95536 | .31206 | .95006 | .32859 | .94447 | 49 |
| 12 | .26219 | .96502 | .27899 | .96029 | .29571 | .95528 | .31233 | .94997 | .32887 | .94438 | 48 |
| 13 | .26247 | .96494 | .27927 | .96021 | .29599 | .95519 | .31261 | .94988 | .32914 | .94428 | 47 |
| 14 | .26275 | .96486 | .27955 | .96013 | .29626 | .95511 | .31289 | .94979 | .32942 | .94418 | 46 |
| 15 | .26303 | .96479 | .27983 | .96005 | .29654 | .95502 | .31316 | .94970 | .32969 | .94409 | 45 |
| 16 | .26331 | .96471 | .28011 | .95997 | .29682 | .95493 | .31344 | .94961 | .32997 | .94399 | 44 |
| 17 | .26359 | .96463 | .28039 | .95989 | .29710 | .95485 | .31372 | .94952 | .33024 | .94390 | 43 |
| 18 | .26387 | .96456 | .28067 | .95981 | .29737 | .95476 | .31399 | .94943 | .33051 | .94380 | 42 |
| 19 | .26415 | .96448 | .28095 | .95972 | .29765 | .95467 | .31427 | .94933 | .33079 | .94370 | 41 |
| 20 | .26443 | .96440 | .28123 | .95964 | .29793 | .95459 | .31454 | .94924 | .33106 | .94361 | 40 |
| 21 | .26471 | .96433 | .28150 | .95956 | .29821 | .95450 | .31482 | .94915 | .33134 | .94351 | 39 |
| 22 | .26500 | .96425 | .28178 | .95948 | .29849 | .95441 | .31510 | .94906 | .33161 | .94342 | 38 |
| 23 | .26528 | .96417 | .28206 | .95940 | .29876 | .95433 | .31537 | .94897 | .33189 | .94332 | 37 |
| 24 | .26556 | .96410 | .28234 | .95931 | .29904 | .95424 | .31565 | .94888 | .33216 | .94322 | 36 |
| 25 | .26584 | .96402 | .28262 | .95923 | .29932 | .95415 | .31593 | .94878 | .33244 | .94313 | 35 |
| 26 | .26612 | .96394 | .28290 | .95915 | .29960 | .95407 | .31620 | .94869 | .33271 | .94303 | 34 |
| 27 | .26640 | .96386 | .28318 | .95907 | .29987 | .95398 | .31648 | .94860 | .33298 | .94293 | 33 |
| 28 | .26668 | .96379 | .28346 | .95898 | .30015 | .95389 | .31675 | .94851 | .33326 | .94284 | 32 |
| 29 | .26696 | .96371 | .28374 | .95890 | .30043 | .95380 | .31703 | .94842 | .33353 | .94274 | 31 |
| 30 | .26724 | .96363 | .28402 | .95882 | .30071 | .95372 | .31730 | .94832 | .33381 | .94264 | 30 |
| 31 | .26752 | .96355 | .28429 | .95874 | .30098 | .95363 | .31758 | .94823 | .33408 | .94254 | 29 |
| 32 | .26780 | .96347 | .28457 | .95865 | .30126 | .95354 | .31786 | .94814 | .33436 | .94245 | 28 |
| 33 | .26808 | .96340 | .28485 | .95857 | .30154 | .95345 | .31813 | .94805 | .33463 | .94235 | 27 |
| 34 | .26836 | .96332 | .28513 | .95849 | .30182 | .95337 | .31841 | .94795 | .33490 | .94225 | 26 |
| 35 | .26864 | .96324 | .28541 | .95841 | .30209 | .95328 | .31868 | .94786 | .33518 | .94215 | 25 |
| 36 | .26892 | .96316 | .28569 | .95832 | .30237 | .95319 | .31896 | .94777 | .33545 | .94206 | 24 |
| 37 | .26920 | .96308 | .28597 | .95824 | .30265 | .95310 | .31923 | .94768 | .33573 | .94196 | 23 |
| 38 | .26948 | .96301 | .28625 | .95816 | .30292 | .95301 | .31951 | .94758 | .33600 | .94186 | 22 |
| 39 | .26976 | .96293 | .28652 | .95807 | .30320 | .95293 | .31979 | .94749 | .33627 | .94176 | 21 |
| 40 | .27004 | .96285 | .28680 | .95799 | .30348 | .95284 | .32006 | .94740 | .33655 | .94167 | 20 |
| 41 | .27032 | .96277 | .28708 | .95791 | .30376 | .95275 | .32034 | .94730 | .33682 | .94157 | 19 |
| 42 | .27060 | .96269 | .28736 | .95782 | .30403 | .95266 | .32061 | .94721 | .33710 | .94147 | 18 |
| 43 | .27088 | .96261 | .28764 | .95774 | .30431 | .95257 | .32089 | .94712 | .33737 | .94137 | 17 |
| 44 | .27116 | .96253 | .28792 | .95766 | .30459 | .95248 | .32116 | .94702 | .33764 | .94127 | 16 |
| 45 | .27144 | .96246 | .28820 | .95757 | .30486 | .95240 | .32144 | .94693 | .33792 | .94118 | 15 |
| 46 | .27172 | .96238 | .28847 | .95749 | .30514 | .95231 | .32171 | .94684 | .33819 | .94108 | 14 |
| 47 | .27200 | .96230 | .28875 | .95740 | .30542 | .95222 | .32199 | .94674 | .33846 | .94098 | 13 |
| 48 | .27228 | .96222 | .28903 | .95732 | .30570 | .95213 | .32227 | .94665 | .33874 | .94088 | 12 |
| 49 | .27256 | .96214 | .28931 | .95724 | .30597 | .95204 | .32254 | .94656 | .33901 | .94078 | 11 |
| 50 | .27284 | .96206 | .28959 | .95715 | .30625 | .95195 | .32282 | .94646 | .33929 | .94068 | 10 |
| 51 | .27312 | .96198 | .28987 | .95707 | .30653 | .95186 | .32309 | .94637 | .33956 | .94058 | 9 |
| 52 | .27340 | .96190 | .29015 | .95698 | .30680 | .95177 | .32337 | .94627 | .33983 | .94049 | 8 |
| 53 | .27368 | .96182 | .29042 | .95690 | .30708 | .95168 | .32364 | .94618 | .34011 | .94039 | 7 |
| 54 | .27396 | .96174 | .29070 | .95681 | .30736 | .95159 | .32392 | .94609 | .34038 | .94029 | 6 |
| 55 | .27424 | .96166 | .29098 | .95673 | .30763 | .95150 | .32419 | .94599 | .34065 | .94019 | 5 |
| 56 | .27452 | .96158 | .29126 | .95664 | .30791 | .95142 | .32447 | .94590 | .34093 | .94009 | 4 |
| 57 | .27480 | .96150 | .29154 | .95656 | .30819 | .95133 | .32474 | .94580 | .34120 | .93999 | 3 |
| 58 | .27508 | .96142 | .29182 | .95647 | .30846 | .95124 | .32502 | .94571 | .34147 | .93989 | 2 |
| 59 | .27536 | .96134 | .29209 | .95639 | .30874 | .95115 | .32529 | .94561 | .34175 | .93979 | 1 |
| 60 | .27564 | .96126 | .29237 | .95630 | .30902 | .95106 | .32557 | .94552 | .34202 | .93969 | 0 |
| | Cosin | Sine | Cosin | Sine | Cosin | Sine | Cosin | Sine | Cosin | Sine | |
| ′ | 74° | | 73° | | 72° | | 71° | | 70° | | ′ |

## NATURAL SINES AND COSINES.

| ′ | 20° Sine | 20° Cosin | 21° Sine | 21° Cosin | 22° Sine | 22° Cosin | 23° Sine | 23° Cosin | 24° Sine | 24° Cosin | ′ |
|---|---|---|---|---|---|---|---|---|---|---|---|
| 0 | .34202 | .93969 | .35837 | .93358 | .37461 | .92718 | .39073 | .92050 | .40674 | .91355 | 60 |
| 1 | .34229 | .93959 | .35864 | .93348 | .37488 | .92707 | .39100 | .92039 | .40700 | .91343 | 59 |
| 2 | .34257 | .93949 | .35891 | .93337 | .37515 | .92697 | .39127 | .92028 | .40727 | .91331 | 58 |
| 3 | .34284 | .93939 | .35918 | .93327 | .37542 | .92686 | .39153 | .92016 | .40753 | .91319 | 57 |
| 4 | .34311 | .93929 | .35945 | .93316 | .37569 | .92675 | .39180 | .92005 | .40780 | .91307 | 56 |
| 5 | .34339 | .93919 | .35973 | .93306 | .37595 | .92664 | .39207 | .91994 | .40806 | .91295 | 55 |
| 6 | .34366 | .93909 | .36000 | .93295 | .37622 | .92653 | .39234 | .91982 | .40833 | .91283 | 54 |
| 7 | .34393 | .93899 | .36027 | .93285 | .37649 | .92642 | .39260 | .91971 | .40860 | .91272 | 53 |
| 8 | .34421 | .93889 | .36054 | .93274 | .37676 | .92631 | .39287 | .91959 | .40886 | .91260 | 52 |
| 9 | .34448 | .93879 | .36081 | .93264 | .37703 | .92620 | .39314 | .91948 | .40913 | .91248 | 51 |
| 10 | .34475 | .93869 | .36108 | .93253 | .37730 | .92609 | .39341 | .91936 | .40939 | .91236 | 50 |
| 11 | .34503 | .93859 | .36135 | .93243 | .37757 | .92598 | .39367 | .91925 | .40966 | .91224 | 49 |
| 12 | .34530 | .93849 | .36162 | .93232 | .37784 | .92587 | .39394 | .91914 | .40992 | .91212 | 48 |
| 13 | .34557 | .93839 | .36190 | .93222 | .37811 | .92576 | .39421 | .91902 | .41019 | .91200 | 47 |
| 14 | .34584 | .93829 | .36217 | .93211 | .37838 | .92565 | .39448 | .91891 | .41045 | .91188 | 46 |
| 15 | .34612 | .93819 | .36244 | .93201 | .37865 | .92554 | .39474 | .91879 | .41072 | .91176 | 45 |
| 16 | .34639 | .93809 | .36271 | .93190 | .37892 | .92543 | .39501 | .91868 | .41098 | .91164 | 44 |
| 17 | .34666 | .93799 | .36298 | .93180 | .37919 | .92532 | .39528 | .91856 | .41125 | .91152 | 43 |
| 18 | .34694 | .93789 | .36325 | .93169 | .37946 | .92521 | .39555 | .91845 | .41151 | .91140 | 42 |
| 19 | .34721 | .93779 | .36352 | .93159 | .37973 | .92510 | .39581 | .91833 | .41178 | .91128 | 41 |
| 20 | .34748 | .93769 | .36379 | .93148 | .37999 | .92499 | .39608 | .91822 | .41204 | .91116 | 40 |
| 21 | .34775 | .93759 | .36406 | .93137 | .38026 | .92488 | .39635 | .91810 | .41231 | .91104 | 39 |
| 22 | .34803 | .93748 | .36434 | .93127 | .38053 | .92477 | .39661 | .91799 | .41257 | .91092 | 38 |
| 23 | .34830 | .93738 | .36461 | .93116 | .38080 | .92466 | .39688 | .91787 | .41284 | .91080 | 37 |
| 24 | .34857 | .93728 | .36488 | .93106 | .38107 | .92455 | .39715 | .91775 | .41310 | .91068 | 36 |
| 25 | .34884 | .93718 | .36515 | .93095 | .38134 | .92444 | .39741 | .91764 | .41337 | .91056 | 35 |
| 26 | .34912 | .93708 | .36542 | .93084 | .38161 | .92432 | .39768 | .91752 | .41363 | .91044 | 34 |
| 27 | .34939 | .93698 | .36569 | .93074 | .38188 | .92421 | .39795 | .91741 | .41390 | .91032 | 33 |
| 28 | .34966 | .93688 | .36596 | .93063 | .38215 | .92410 | .39822 | .91729 | .41416 | .91020 | 32 |
| 29 | .34993 | .93677 | .36623 | .93052 | .38241 | .92399 | .39848 | .91718 | .41443 | .91008 | 31 |
| 30 | .35021 | .93667 | .36650 | .93042 | .38268 | .92388 | .39875 | .91706 | .41469 | .90996 | 30 |
| 31 | .35048 | .93657 | .36677 | .93031 | .38295 | .92377 | .39902 | .91694 | .41496 | .90984 | 29 |
| 32 | .35075 | .93647 | .36704 | .93020 | .38322 | .92366 | .39928 | .91683 | .41522 | .90972 | 28 |
| 33 | .35102 | .93637 | .36731 | .93010 | .38349 | .92355 | .39955 | .91671 | .41549 | .90960 | 27 |
| 34 | .35130 | .93626 | .36758 | .92999 | .38376 | .92343 | .39982 | .91660 | .41575 | .90948 | 26 |
| 35 | .35157 | .93616 | .36785 | .92988 | .38403 | .92332 | .40008 | .91648 | .41602 | .90936 | 25 |
| 36 | .35184 | .93606 | .36812 | .92978 | .38430 | .92321 | .40035 | .91636 | .41628 | .90924 | 24 |
| 37 | .35211 | .93596 | .36839 | .92967 | .38456 | .92310 | .40062 | .91625 | .41655 | .90911 | 23 |
| 38 | .35239 | .93585 | .36867 | .92956 | .38483 | .92299 | .40088 | .91613 | .41681 | .90899 | 22 |
| 39 | .35266 | .93575 | .36894 | .92945 | .38510 | .92287 | .40115 | .91601 | .41707 | .90887 | 21 |
| 40 | .35293 | .93565 | .36921 | .92935 | .38537 | .92276 | .40141 | .91590 | .41734 | .90875 | 20 |
| 41 | .35320 | .93555 | .36948 | .92924 | .38564 | .92265 | .40168 | .91578 | .41760 | .90863 | 19 |
| 42 | .35347 | .93544 | .36975 | .92913 | .38591 | .92254 | .40195 | .91566 | .41787 | .90851 | 18 |
| 43 | .35375 | .93534 | .37002 | .92902 | .38617 | .92243 | .40221 | .91555 | .41813 | .90839 | 17 |
| 44 | .35402 | .93524 | .37029 | .92892 | .38644 | .92231 | .40248 | .91543 | .41840 | .90826 | 16 |
| 45 | .35429 | .93514 | .37056 | .92881 | .38671 | .92220 | .40275 | .91531 | .41866 | .90814 | 15 |
| 46 | .35456 | .93503 | .37083 | .92870 | .38698 | .92209 | .40301 | .91519 | .41892 | .90802 | 14 |
| 47 | .35484 | .93493 | .37110 | .92859 | .38725 | .92198 | .40328 | .91508 | .41919 | .90790 | 13 |
| 48 | .35511 | .93483 | .37137 | .92849 | .38752 | .92186 | .40355 | .91496 | .41945 | .90778 | 12 |
| 49 | .35538 | .93472 | .37164 | .92838 | .38778 | .92175 | .40381 | .91484 | .41972 | .90766 | 11 |
| 50 | .35565 | .93462 | .37191 | .92827 | .38805 | .92164 | .40408 | .91472 | .41998 | .90753 | 10 |
| 51 | .35592 | .93452 | .37218 | .92816 | .38832 | .92152 | .40434 | .91461 | .42024 | .90741 | 9 |
| 52 | .35619 | .93441 | .37245 | .92805 | .38859 | .92141 | .40461 | .91449 | .42051 | .90729 | 8 |
| 53 | .35647 | .93431 | .37272 | .92794 | .38886 | .92130 | .40488 | .91437 | .42077 | .90717 | 7 |
| 54 | .35674 | .93420 | .37299 | .92784 | .38912 | .92119 | .40514 | .91425 | .42104 | .90704 | 6 |
| 55 | .35701 | .93410 | .37326 | .92773 | .38939 | .92107 | .40541 | .91414 | .42130 | .90692 | 5 |
| 56 | .35728 | .93400 | .37353 | .92762 | .38966 | .92096 | .40567 | .91402 | .42156 | .90680 | 4 |
| 57 | .35755 | .93389 | .37380 | .92751 | .38993 | .92085 | .40594 | .91390 | .42183 | .90668 | 3 |
| 58 | .35782 | .93379 | .37407 | .92740 | .39020 | .92073 | .40621 | .91378 | .42209 | .90655 | 2 |
| 59 | .35810 | .93368 | .37434 | .92729 | .39046 | .92062 | .40647 | .91366 | .42235 | .90643 | 1 |
| 60 | .35837 | .93358 | .37461 | .92718 | .39073 | .92050 | .40674 | .91355 | .42262 | .90631 | 0 |
| ′ | Cosin | Sine | Cosin | Sine | Cosin | Sine | Cosin | Sine | Cosin | Sine | ′ |
|  | 69° |  | 68° |  | 67° |  | 66° |  | 65° |  |  |

## NATURAL SINES AND COSINES. 117

| ′ | 25° Sine | Cosin | 26° Sine | Cosin | 27° Sine | Cosin | 28° Sine | Cosin | 29° Sine | Cosin | ′ |
|---|---|---|---|---|---|---|---|---|---|---|---|
| 0 | .42262 | .90631 | .43837 | .89879 | .45399 | .89101 | .46947 | .88295 | .48481 | .87462 | 60 |
| 1 | .42288 | .90618 | .43863 | .89867 | .45425 | .89087 | .46973 | .88281 | .48506 | .87448 | 59 |
| 2 | .42315 | .90606 | .43889 | .89854 | .45451 | .89074 | .46999 | .88267 | .48532 | .87434 | 58 |
| 3 | .42341 | .90594 | .43916 | .89841 | .45477 | .89061 | .47024 | .88254 | .48557 | .87420 | 57 |
| 4 | .42367 | .90582 | .43942 | .89828 | .45503 | .89048 | .47050 | .88240 | .48583 | .87406 | 56 |
| 5 | .42394 | .90569 | .43968 | .89816 | .45529 | .89035 | .47076 | .88226 | .48608 | .87391 | 55 |
| 6 | .42420 | .90557 | .43994 | .89803 | .45554 | .89021 | .47101 | .88213 | .48634 | .87377 | 54 |
| 7 | .42446 | .90545 | .44020 | .89790 | .45580 | .89008 | .47127 | .88199 | .48659 | .87363 | 53 |
| 8 | .42473 | .90532 | .44046 | .89777 | .45606 | .88995 | .47153 | .88185 | .48684 | .87349 | 52 |
| 9 | .42499 | .90520 | .44072 | .89764 | .45632 | .88981 | .47178 | .88172 | .48710 | .87335 | 51 |
| 10 | .42525 | .90507 | .44098 | .89752 | .45658 | .88968 | .47204 | .88158 | .48735 | .87321 | 50 |
| 11 | .42552 | .90495 | .44124 | .89739 | .45684 | .88955 | .47229 | .88144 | .48761 | .87306 | 49 |
| 12 | .42578 | .90483 | .44151 | .89726 | .45710 | .88942 | .47255 | .88130 | .48786 | .87292 | 48 |
| 13 | .42604 | .90470 | .44177 | .89713 | .45736 | .88928 | .47281 | .88117 | .48811 | .87278 | 47 |
| 14 | .42631 | .90458 | .44203 | .89700 | .45762 | .88915 | .47306 | .88103 | .48837 | .87264 | 46 |
| 15 | .42657 | .90446 | .44229 | .89687 | .45787 | .88902 | .47332 | .88089 | .48862 | .87250 | 45 |
| 16 | .42683 | .90433 | .44255 | .89674 | .45813 | .88888 | .47358 | .88075 | .48888 | .87235 | 44 |
| 17 | .42709 | .90421 | .44281 | .89662 | .45839 | .88875 | .47383 | .88062 | .48913 | .87221 | 43 |
| 18 | .42736 | .90408 | .44307 | .89649 | .45865 | .88862 | .47409 | .88048 | .48938 | .87207 | 42 |
| 19 | .42762 | .90396 | .44333 | .89636 | .45891 | .88848 | .47434 | .88034 | .48964 | .87193 | 41 |
| 20 | .42788 | .90383 | .44359 | .89623 | .45917 | .88835 | .47460 | .88020 | .48989 | .87178 | 40 |
| 21 | .42815 | .90371 | .44385 | .89610 | .45942 | .88822 | .47486 | .88006 | .49014 | .87164 | 39 |
| 22 | .42841 | .90358 | .44411 | .89597 | .45968 | .88808 | .47511 | .87993 | .49040 | .87150 | 38 |
| 23 | .42867 | .90346 | .44437 | .89584 | .45994 | .88795 | .47537 | .87979 | .49065 | .87136 | 37 |
| 24 | .42894 | .90334 | .44464 | .89571 | .46020 | .88782 | .47562 | .87965 | .49090 | .87121 | 36 |
| 25 | .42920 | .90321 | .44490 | .89558 | .46046 | .88768 | .47588 | .87951 | .49116 | .87107 | 35 |
| 26 | .42946 | .90309 | .44516 | .89545 | .46072 | .88755 | .47614 | .87937 | .49141 | .87093 | 34 |
| 27 | .42972 | .90296 | .44542 | .89532 | .46097 | .88741 | .47639 | .87923 | .49166 | .87079 | 33 |
| 28 | .42999 | .90284 | .44568 | .89519 | .46123 | .88728 | .47665 | .87909 | .49192 | .87064 | 32 |
| 29 | .43025 | .90271 | .44594 | .89506 | .46149 | .88715 | .47690 | .87896 | .49217 | .87050 | 31 |
| 30 | .43051 | .90259 | .44620 | .89493 | .46175 | .88701 | .47716 | .87882 | .49242 | .87036 | 30 |
| 31 | .43077 | .90246 | .44646 | .89480 | .46201 | .88688 | .47741 | .87868 | .49268 | .87021 | 29 |
| 32 | .43104 | .90233 | .44672 | .89467 | .46226 | .88674 | .47767 | .87854 | .49293 | .87007 | 28 |
| 33 | .43130 | .90221 | .44698 | .89454 | .46252 | .88661 | .47793 | .87840 | .49318 | .86993 | 27 |
| 34 | .43156 | .90208 | .44724 | .89441 | .46278 | .88647 | .47818 | .87826 | .49344 | .86978 | 26 |
| 35 | .43182 | .90196 | .44750 | .89428 | .46304 | .88634 | .47844 | .87812 | .49369 | .86964 | 25 |
| 36 | .43209 | .90183 | .44776 | .89415 | .46330 | .88620 | .47869 | .87798 | .49394 | .86949 | 24 |
| 37 | .43235 | .90171 | .44802 | .89402 | .46355 | .88607 | .47895 | .87784 | .49419 | .86935 | 23 |
| 38 | .43261 | .90158 | .44828 | .89389 | .46381 | .88593 | .47920 | .87770 | .49445 | .86921 | 22 |
| 39 | .43287 | .90146 | .44854 | .89376 | .46407 | .88580 | .47946 | .87756 | .49470 | .86906 | 21 |
| 40 | .43313 | .90133 | .44880 | .89363 | .46433 | .88566 | .47971 | .87743 | .49495 | .86892 | 20 |
| 41 | .43340 | .90120 | .44906 | .89350 | .46458 | .88553 | .47997 | .87729 | .49521 | .86878 | 19 |
| 42 | .43366 | .90108 | .44932 | .89337 | .46484 | .88539 | .48022 | .87715 | .49546 | .86863 | 18 |
| 43 | .43392 | .90095 | .44958 | .89324 | .46510 | .88526 | .48048 | .87701 | .49571 | .86849 | 17 |
| 44 | .43418 | .90082 | .44984 | .89311 | .46536 | .88512 | .48073 | .87687 | .49596 | .86834 | 16 |
| 45 | .43445 | .90070 | .45010 | .89298 | .46561 | .88499 | .48099 | .87673 | .49622 | .86820 | 15 |
| 46 | .43471 | .90057 | .45036 | .89285 | .46587 | .88485 | .48124 | .87659 | .49647 | .86805 | 14 |
| 47 | .43497 | .90045 | .45062 | .89272 | .46613 | .88472 | .48150 | .87645 | .49672 | .86791 | 13 |
| 48 | .43523 | .90032 | .45088 | .89259 | .46639 | .88458 | .48175 | .87631 | .49697 | .86777 | 12 |
| 49 | .43549 | .90019 | .45114 | .89245 | .46664 | .88445 | .48201 | .87617 | .49723 | .86762 | 11 |
| 50 | .43575 | .90007 | .45140 | .89232 | .46690 | .88431 | .48226 | .87603 | .49748 | .86748 | 10 |
| 51 | .43602 | .89994 | .45166 | .89219 | .46716 | .88417 | .48252 | .87589 | .49773 | .86733 | 9 |
| 52 | .43628 | .89981 | .45192 | .89206 | .46742 | .88404 | .48277 | .87575 | .49798 | .86719 | 8 |
| 53 | .43654 | .89968 | .45218 | .89193 | .46767 | .88390 | .48303 | .87561 | .49824 | .86704 | 7 |
| 54 | .43680 | .89956 | .45243 | .89180 | .46793 | .88377 | .48328 | .87546 | .49849 | .86690 | 6 |
| 55 | .43706 | .89943 | .45269 | .89167 | .46819 | .88363 | .48354 | .87532 | .49874 | .86675 | 5 |
| 56 | .43733 | .89930 | .45295 | .89153 | .46844 | .88349 | .48379 | .87518 | .49899 | .86661 | 4 |
| 57 | .43759 | .89918 | .45321 | .89140 | .46870 | .88336 | .48405 | .87504 | .49924 | .86646 | 3 |
| 58 | .43785 | .89905 | .45347 | .89127 | .46896 | .88322 | .48430 | .87490 | .49950 | .86632 | 2 |
| 59 | .43811 | .89892 | .45373 | .89114 | .46921 | .88308 | .48456 | .87476 | .49975 | .86617 | 1 |
| 60 | .43837 | .89879 | .45399 | .89101 | .46947 | .88295 | .48481 | .87462 | .50000 | .86603 | 0 |
| ′ | Cosin | Sine | Cosin | Sine | Cosin | Sine | Cosin | Sine | Cosin | Sine | ′ |
|  | 64° | | 63° | | 62° | | 61° | | 60° | | |

| ′ | 30° | | 31° | | 32° | | 33° | | 34° | | ′ |
|---|---|---|---|---|---|---|---|---|---|---|---|
| | Sine | Cosin | Sine | Cosin | Sine | Cosin | Sine | Cosin | Sine | Cosin | |
| 0 | .50000 | .86603 | .51504 | .85717 | .52992 | .84805 | .54464 | .83867 | .55919 | .82904 | 60 |
| 1 | .50025 | .86588 | .51529 | .85702 | .53017 | .84789 | .54488 | .83851 | .55943 | .82887 | 59 |
| 2 | .50050 | .86573 | .51554 | .85687 | .53041 | .84774 | .54513 | .83835 | .55968 | .82871 | 58 |
| 3 | .50076 | .86559 | .51579 | .85672 | .53066 | .84759 | .54537 | .83819 | .55992 | .82855 | 57 |
| 4 | .50101 | .86544 | .51604 | .85657 | .53091 | .84743 | .54561 | .83804 | .56016 | .82839 | 56 |
| 5 | .50126 | .86530 | .51628 | .85642 | .53115 | .84728 | .54586 | .83788 | .56040 | .82822 | 55 |
| 6 | .50151 | .86515 | .51653 | .85627 | .53140 | .84712 | .54610 | .83772 | .56064 | .82806 | 54 |
| 7 | .50176 | .86501 | .51678 | .85612 | .53164 | .84697 | .54635 | .83756 | .56088 | .82790 | 53 |
| 8 | .50201 | .86486 | .51703 | .85597 | .53189 | .84681 | .54659 | .83740 | .56112 | .82773 | 52 |
| 9 | .50227 | .86471 | .51728 | .85582 | .53214 | .84666 | .54683 | .83724 | .56136 | .82757 | 51 |
| 10 | .50252 | .86457 | .51753 | .85567 | .53238 | .84650 | .54708 | .83708 | .56160 | .82741 | 50 |
| 11 | .50277 | .86442 | .51778 | .85551 | .53263 | .84635 | .54732 | .83692 | .56184 | .82724 | 49 |
| 12 | .50302 | .86427 | .51803 | .85536 | .53288 | .84619 | .54756 | .83676 | .56208 | .82708 | 48 |
| 13 | .50327 | .86413 | .51828 | .85521 | .53312 | .84604 | .54781 | .83660 | .56232 | .82692 | 47 |
| 14 | .50352 | .86398 | .51852 | .85506 | .53337 | .84588 | .54805 | .83645 | .56256 | .82675 | 46 |
| 15 | .50377 | .86384 | .51877 | .85491 | .53361 | .84573 | .54829 | .83629 | .56280 | .82659 | 45 |
| 16 | .50403 | .86369 | .51902 | .85476 | .53386 | .84557 | .54854 | .83613 | .56305 | .82643 | 44 |
| 17 | .50428 | .86354 | .51927 | .85461 | .53411 | .84542 | .54878 | .83597 | .56329 | .82626 | 43 |
| 18 | .50453 | .86340 | .51952 | .85446 | .53435 | .84526 | .54902 | .83581 | .56353 | .82610 | 42 |
| 19 | .50478 | .86325 | .51977 | .85431 | .53460 | .84511 | .54927 | .83565 | .56377 | .82593 | 41 |
| 20 | .50503 | .86310 | .52002 | .85416 | .53484 | .84495 | .54951 | .83549 | .56401 | .82577 | 40 |
| 21 | .50528 | .86295 | .52026 | .85401 | .53509 | .84480 | .54975 | .83533 | .56425 | .82561 | 39 |
| 22 | .50553 | .86281 | .52051 | .85385 | .53534 | .84464 | .54999 | .83517 | .56449 | .82544 | 38 |
| 23 | .50578 | .86266 | .52076 | .85370 | .53558 | .84448 | .55024 | .83501 | .56473 | .82528 | 37 |
| 24 | .50603 | .86251 | .52101 | .85355 | .53583 | .84433 | .55048 | .83485 | .56497 | .82511 | 36 |
| 25 | .50628 | .86237 | .52126 | .85340 | .53607 | .84417 | .55072 | .83469 | .56521 | .82495 | 35 |
| 26 | .50654 | .86222 | .52151 | .85325 | .53632 | .84402 | .55097 | .83453 | .56545 | .82478 | 34 |
| 27 | .50679 | .86207 | .52175 | .85310 | .53656 | .84386 | .55121 | .83437 | .56569 | .82462 | 33 |
| 28 | .50704 | .86192 | .52200 | .85294 | .53681 | .84370 | .55145 | .83421 | .56593 | .82446 | 32 |
| 29 | .50729 | .86178 | .52225 | .85279 | .53705 | .84355 | .55169 | .83405 | .56617 | .82429 | 31 |
| 30 | .50754 | .86163 | .52250 | .85264 | .53730 | .84339 | .55194 | .83389 | .56641 | .82413 | 30 |
| 31 | .50779 | .86148 | .52275 | .85249 | .53754 | .84324 | .55218 | .83373 | .56665 | .82396 | 29 |
| 32 | .50804 | .86133 | .52299 | .85234 | .53779 | .84308 | .55242 | .83356 | .56689 | .82380 | 28 |
| 33 | .50829 | .86119 | .52324 | .85218 | .53804 | .84292 | .55266 | .83340 | .56713 | .82363 | 27 |
| 34 | .50854 | .86104 | .52349 | .85203 | .53828 | .84277 | .55291 | .83324 | .56736 | .82347 | 26 |
| 35 | .50879 | .86089 | .52374 | .85188 | .53853 | .84261 | .55315 | .83308 | .56760 | .82330 | 25 |
| 36 | .50904 | .86074 | .52399 | .85173 | .53877 | .84245 | .55339 | .83292 | .56784 | .82314 | 24 |
| 37 | .50929 | .86059 | .52423 | .85157 | .53902 | .84230 | .55363 | .83276 | .56808 | .82297 | 23 |
| 38 | .50954 | .86045 | .52448 | .85142 | .53926 | .84214 | .55388 | .83260 | .56832 | .82281 | 22 |
| 39 | .50979 | .86030 | .52473 | .85127 | .53951 | .84198 | .55412 | .83244 | .56856 | .82264 | 21 |
| 40 | .51004 | .86015 | .52498 | .85112 | .53975 | .84182 | .55436 | .83228 | .56880 | .82248 | 20 |
| 41 | .51029 | .86000 | .52522 | .85096 | .54000 | .84167 | .55460 | .83212 | .56904 | .82231 | 19 |
| 42 | .51054 | .85985 | .52547 | .85081 | .54024 | .84151 | .55484 | .83195 | .56928 | .82214 | 18 |
| 43 | .51079 | .85970 | .52572 | .85066 | .54049 | .84135 | .55509 | .83179 | .56952 | .82198 | 17 |
| 44 | .51104 | .85956 | .52597 | .85051 | .54073 | .84120 | .55533 | .83163 | .56976 | .82181 | 16 |
| 45 | .51129 | .85941 | .52621 | .85035 | .54097 | .84104 | .55557 | .83147 | .57000 | .82165 | 15 |
| 46 | .51154 | .85926 | .52646 | .85020 | .54122 | .84088 | .55581 | .83131 | .57024 | .82148 | 14 |
| 47 | .51179 | .85911 | .52671 | .85005 | .54146 | .84072 | .55605 | .83115 | .57047 | .82132 | 13 |
| 48 | .51204 | .85896 | .52696 | .84989 | .54171 | .84057 | .55630 | .83098 | .57071 | .82115 | 12 |
| 49 | .51229 | .85881 | .52720 | .84974 | .54195 | .84041 | .55654 | .83082 | .57095 | .82098 | 11 |
| 50 | .51254 | .85866 | .52745 | .84959 | .54220 | .84025 | .55678 | .83066 | .57119 | .82082 | 10 |
| 51 | .51279 | .85851 | .52770 | .84943 | .54244 | .84009 | .55702 | .83050 | .57143 | .82065 | 9 |
| 52 | .51304 | .85836 | .52794 | .84928 | .54269 | .83994 | .55726 | .83034 | .57167 | .82048 | 8 |
| 53 | .51329 | .85821 | .52819 | .84913 | .54293 | .83978 | .55750 | .83017 | .57191 | .82032 | 7 |
| 54 | .51354 | .85806 | .52844 | .84897 | .54317 | .83962 | .55775 | .83001 | .57215 | .82015 | 6 |
| 55 | .51379 | .85792 | .52869 | .84882 | .54342 | .83946 | .55799 | .82985 | .57238 | .81999 | 5 |
| 56 | .51404 | .85777 | .52893 | .84866 | .54366 | .83930 | .55823 | .82969 | .57262 | .81982 | 4 |
| 57 | .51429 | .85762 | .52918 | .84851 | .54391 | .83915 | .55847 | .82953 | .57286 | .81965 | 3 |
| 58 | .51454 | .85747 | .52943 | .84836 | .54415 | .83899 | .55871 | .82936 | .57310 | .81949 | 2 |
| 59 | .51479 | .85732 | .52967 | .84820 | .54440 | .83883 | .55895 | .82920 | .57334 | .81932 | 1 |
| 60 | .51504 | .85717 | .52992 | .84805 | .54464 | .83867 | .55919 | .82904 | .57358 | .81915 | 0 |
| ′ | Cosin | Sine | Cosin | Sine | Cosin | Sine | Cosin | Sine | Cosin | Sine | ′ |
| | 59° | | 58° | | 57° | | 56° | | 55° | | |

## NATURAL SINES AND COSINES. 119

| ′ | 35° | | 36° | | 37° | | 38° | | 39° | | ′ |
|---|---|---|---|---|---|---|---|---|---|---|---|
| | Sine | Cosin | Sine | Cosin | Sine | Cosin | Sine | Cosin | Sine | Cosin | |
| 0 | .57358 | .81915 | .58779 | .80902 | .60182 | .79864 | .61566 | .78801 | .62932 | .77715 | 60 |
| 1 | .57381 | .81899 | .58802 | .80885 | .60205 | .79846 | .61589 | .78783 | .62955 | .77696 | 59 |
| 2 | .57405 | .81882 | .58826 | .80867 | .60228 | .79829 | .61612 | .78765 | .62977 | .77678 | 58 |
| 3 | .57429 | .81865 | .58849 | .80850 | .60251 | .79811 | .61635 | .78747 | .63000 | .77660 | 57 |
| 4 | .57453 | .81848 | .58873 | .80833 | .60274 | .79793 | .61658 | .78729 | .63022 | .77641 | 56 |
| 5 | .57477 | .81832 | .58896 | .80816 | .60298 | .79776 | .61681 | .78711 | .63045 | .77623 | 55 |
| 6 | .57501 | .81815 | .58920 | .80799 | .60321 | .79758 | .61704 | .78694 | .63068 | .77605 | 54 |
| 7 | .57524 | .81798 | .58943 | .80782 | .60344 | .79741 | .61726 | .78676 | .63090 | .77586 | 53 |
| 8 | .57548 | .81782 | .58967 | .80765 | .60367 | .79723 | .61749 | .78658 | .63113 | .77568 | 52 |
| 9 | .57572 | .81765 | .58990 | .80748 | .60390 | .79706 | .61772 | .78640 | .63135 | .77550 | 51 |
| 10 | .57596 | .81748 | .59014 | .80730 | .60414 | .79688 | .61795 | .78622 | .63158 | .77531 | 50 |
| 11 | .57619 | .81731 | .59037 | .80713 | .60437 | .79671 | .61818 | .78604 | .63180 | .77513 | 49 |
| 12 | .57643 | .81714 | .59061 | .80696 | .60460 | .79653 | .61841 | .78586 | .63203 | .77494 | 48 |
| 13 | .57667 | .81698 | .59084 | .80679 | .60483 | .79635 | .61864 | .78568 | .63225 | .77476 | 47 |
| 14 | .57691 | .81681 | .59108 | .80662 | .60506 | .79618 | .61887 | .78550 | .63248 | .77458 | 46 |
| 15 | .57715 | .81664 | .59131 | .80644 | .60529 | .79600 | .61909 | .78532 | .63271 | .77439 | 45 |
| 16 | .57738 | .81647 | .59154 | .80627 | .60553 | .79583 | .61932 | .78514 | .63293 | .77421 | 44 |
| 17 | .57762 | .81631 | .59178 | .80610 | .60576 | .79565 | .61955 | .78496 | .63316 | .77402 | 43 |
| 18 | .57786 | .81614 | .59201 | .80593 | .60599 | .79547 | .61978 | .78478 | .63338 | .77384 | 42 |
| 19 | .57810 | .81597 | .59225 | .80576 | .60622 | .79530 | .62001 | .78460 | .63361 | .77366 | 41 |
| 20 | .57833 | .81580 | .59248 | .80558 | .60645 | .79512 | .62024 | .78442 | .63383 | .77347 | 40 |
| 21 | .57857 | .81563 | .59272 | .80541 | .60668 | .79494 | .62046 | .78424 | .63406 | .77329 | 39 |
| 22 | .57881 | .81546 | .59295 | .80524 | .60691 | .79477 | .62069 | .78405 | .63428 | .77310 | 38 |
| 23 | .57904 | .81530 | .59318 | .80507 | .60714 | .79459 | .62092 | .78387 | .63451 | .77292 | 37 |
| 24 | .57928 | .81513 | .59342 | .80489 | .60738 | .79441 | .62115 | .78369 | .63473 | .77273 | 36 |
| 25 | .57952 | .81496 | .59365 | .80472 | .60761 | .79424 | .62138 | .78351 | .63496 | .77255 | 35 |
| 26 | .57976 | .81479 | .59389 | .80455 | .60784 | .79406 | .62160 | .78333 | .63518 | .77236 | 34 |
| 27 | .57999 | .81462 | .59412 | .80438 | .60807 | .79388 | .62183 | .78315 | .63540 | .77218 | 33 |
| 28 | .58023 | .81445 | .59436 | .80420 | .60830 | .79371 | .62206 | .78297 | .63563 | .77199 | 32 |
| 29 | .58047 | .81428 | .59459 | .80403 | .60853 | .79353 | .62229 | .78279 | .63585 | .77181 | 31 |
| 30 | .58070 | .81412 | .59482 | .80386 | .60876 | .79335 | .62251 | .78261 | .63608 | .77162 | 30 |
| 31 | .58094 | .81395 | .59506 | .80368 | .60899 | .79318 | .62274 | .78243 | .63630 | .77144 | 29 |
| 32 | .58118 | .81378 | .59529 | .80351 | .60922 | .79300 | .62297 | .78225 | .63653 | .77125 | 28 |
| 33 | .58141 | .81361 | .59552 | .80334 | .60945 | .79282 | .62320 | .78206 | .63675 | .77107 | 27 |
| 34 | .58165 | .81344 | .59576 | .80316 | .60968 | .79264 | .62342 | .78188 | .63698 | .77088 | 26 |
| 35 | .58189 | .81327 | .59599 | .80299 | .60991 | .79247 | .62365 | .78170 | .63720 | .77070 | 25 |
| 36 | .58212 | .81310 | .59622 | .80282 | .61015 | .79229 | .62388 | .78152 | .63742 | .77051 | 24 |
| 37 | .58236 | .81293 | .59646 | .80264 | .61038 | .79211 | .62411 | .78134 | .63765 | .77033 | 23 |
| 38 | .58260 | .81276 | .59669 | .80247 | .61061 | .79193 | .62433 | .78116 | .63787 | .77014 | 22 |
| 39 | .58283 | .81259 | .59693 | .80230 | .61084 | .79176 | .62456 | .78098 | .63810 | .76996 | 21 |
| 40 | .58307 | .81242 | .59716 | .80212 | .61107 | .79158 | .62479 | .78079 | .63832 | .76977 | 20 |
| 41 | .58330 | .81225 | .59739 | .80195 | .61130 | .79140 | .62502 | .78061 | .63854 | .76959 | 19 |
| 42 | .58354 | .81208 | .59763 | .80178 | .61153 | .79122 | .62524 | .78043 | .63877 | .76940 | 18 |
| 43 | .58378 | .81191 | .59786 | .80160 | .61176 | .79105 | .62547 | .78025 | .63899 | .76921 | 17 |
| 44 | .58401 | .81174 | .59809 | .80143 | .61199 | .79087 | .62570 | .78007 | .63922 | .76903 | 16 |
| 45 | .58425 | .81157 | .59832 | .80125 | .61222 | .79069 | .62592 | .77988 | .63944 | .76884 | 15 |
| 46 | .58449 | .81140 | .59856 | .80108 | .61245 | .79051 | .62615 | .77970 | .63966 | .76866 | 14 |
| 47 | .58472 | .81123 | .59879 | .80091 | .61268 | .79033 | .62638 | .77952 | .63989 | .76847 | 13 |
| 48 | .58496 | .81106 | .59902 | .80073 | .61291 | .79016 | .62660 | .77934 | .64011 | .76828 | 12 |
| 49 | .58519 | .81089 | .59926 | .80056 | .61314 | .78998 | .62683 | .77916 | .64033 | .76810 | 11 |
| 50 | .58543 | .81072 | .59949 | .80038 | .61337 | .78980 | .62706 | .77897 | .64056 | .76791 | 10 |
| 51 | .58567 | .81055 | .59972 | .80021 | .61360 | .78962 | .62728 | .77879 | .64078 | .76772 | 9 |
| 52 | .58590 | .81038 | .59995 | .80003 | .61383 | .78944 | .62751 | .77861 | .64100 | .76754 | 8 |
| 53 | .58614 | .81021 | .60019 | .79986 | .61406 | .78926 | .62774 | .77843 | .64123 | .76735 | 7 |
| 54 | .58637 | .81004 | .60042 | .79968 | .61429 | .78908 | .62796 | .77824 | .64145 | .76717 | 6 |
| 55 | .58661 | .80987 | .60065 | .79951 | .61451 | .78891 | .62819 | .77806 | .64167 | .76698 | 5 |
| 56 | .58684 | .80970 | .60089 | .79934 | .61474 | .78873 | .62842 | .77788 | .64190 | .76679 | 4 |
| 57 | .58708 | .80953 | .60112 | .79916 | .61497 | .78855 | .62864 | .77769 | .64212 | .76661 | 3 |
| 58 | .58731 | .80936 | .60135 | .79899 | .61520 | .78837 | .62887 | .77751 | .64234 | .76642 | 2 |
| 59 | .58755 | .80919 | .60158 | .79881 | .61543 | .78819 | .62909 | .77733 | .64256 | .76623 | 1 |
| 60 | .58779 | .80902 | .60182 | .79864 | .61566 | .78801 | .62932 | .77715 | .64279 | .76604 | 0 |
| | Cosin | Sine | Cosin | Sine | Cosin | Sine | Cosin | Sine | Cosin | Sine | |
| ′ | 54° | | 53° | | 52° | | 51° | | 50° | | ′ |

## NATURAL SINES AND COSINES.

| ′ | 40° Sine | 40° Cosin | 41° Sine | 41° Cosin | 42° Sine | 42° Cosin | 43° Sine | 43° Cosin | 44° Sine | 44° Cosin | ′ |
|---|---|---|---|---|---|---|---|---|---|---|---|
| 0 | .64279 | .76604 | .65606 | .75471 | .66913 | .74314 | .68200 | .73135 | .69466 | .71934 | 60 |
| 1 | .64301 | .76586 | .65628 | .75452 | .66935 | .74295 | .68221 | .73116 | .69487 | .71914 | 59 |
| 2 | .64323 | .76567 | .65650 | .75433 | .66956 | .74276 | .68242 | .73096 | .69508 | .71894 | 58 |
| 3 | .64346 | .76548 | .65672 | .75414 | .66978 | .74256 | .68264 | .73076 | .69529 | .71873 | 57 |
| 4 | .64368 | .76530 | .65694 | .75395 | .66999 | .74237 | .68285 | .73056 | .69549 | .71853 | 56 |
| 5 | .64390 | .76511 | .65716 | .75375 | .67021 | .74217 | .68306 | .73036 | .69570 | .71833 | 55 |
| 6 | .64412 | .76492 | .65738 | .75356 | .67043 | .74198 | .68327 | .73016 | .69591 | .71813 | 54 |
| 7 | .64435 | .76473 | .65759 | .75337 | .67064 | .74178 | .68349 | .72996 | .69612 | .71792 | 53 |
| 8 | .64457 | .76455 | .65781 | .75318 | .67086 | .74159 | .68370 | .72976 | .69633 | .71772 | 52 |
| 9 | .64479 | .76436 | .65803 | .75299 | .67107 | .74139 | .68391 | .72957 | .69654 | .71752 | 51 |
| 10 | .64501 | .76417 | .65825 | .75280 | .67129 | .74120 | .68412 | .72937 | .69675 | .71732 | 50 |
| 11 | .64524 | .76398 | .65847 | .75261 | .67151 | .74100 | .68434 | .72917 | .69696 | .71711 | 49 |
| 12 | .64546 | .76380 | .65869 | .75241 | .67172 | .74080 | .68455 | .72897 | .69717 | .71691 | 48 |
| 13 | .64568 | .76361 | .65891 | .75222 | .67194 | .74061 | .68476 | .72877 | .69737 | .71671 | 47 |
| 14 | .64590 | .76342 | .65913 | .75203 | .67215 | .74041 | .68497 | .72857 | .69758 | .71650 | 46 |
| 15 | .64612 | .76323 | .65935 | .75184 | .67237 | .74022 | .68518 | .72837 | .69779 | .71630 | 45 |
| 16 | .64635 | .76304 | .65956 | .75165 | .67258 | .74002 | .68539 | .72817 | .69800 | .71610 | 44 |
| 17 | .64657 | .76286 | .65978 | .75146 | .67280 | .73983 | .68561 | .72797 | .69821 | .71590 | 43 |
| 18 | .64679 | .76267 | .66000 | .75126 | .67301 | .73963 | .68582 | .72777 | .69842 | .71569 | 42 |
| 19 | .64701 | .76248 | .66022 | .75107 | .67323 | .73944 | .68603 | .72757 | .69862 | .71549 | 41 |
| 20 | .64723 | .76229 | .66044 | .75088 | .67344 | .73924 | .68624 | .72737 | .69883 | .71529 | 40 |
| 21 | .64746 | .76210 | .66066 | .75069 | .67366 | .73904 | .68645 | .72717 | .69904 | .71508 | 39 |
| 22 | .64768 | .76192 | .66088 | .75050 | .67387 | .73885 | .68666 | .72697 | .69925 | .71488 | 38 |
| 23 | .64790 | .76173 | .66109 | .75030 | .67409 | .73865 | .68688 | .72677 | .69946 | .71468 | 37 |
| 24 | .64812 | .76154 | .66131 | .75011 | .67430 | .73846 | .68709 | .72657 | .69966 | .71447 | 36 |
| 25 | .64834 | .76135 | .66153 | .74992 | .67452 | .73826 | .68730 | .72637 | .69987 | .71427 | 35 |
| 26 | .64856 | .76116 | .66175 | .74973 | .67473 | .73806 | .68751 | .72617 | .70008 | .71407 | 34 |
| 27 | .64878 | .76097 | .66197 | .74953 | .67495 | .73787 | .68772 | .72597 | .70029 | .71386 | 33 |
| 28 | .64901 | .76078 | .66218 | .74934 | .67516 | .73767 | .68793 | .72577 | .70049 | .71366 | 32 |
| 29 | .64923 | .76059 | .66240 | .74915 | .67538 | .73747 | .68814 | .72557 | .70070 | .71345 | 31 |
| 30 | .64945 | .76041 | .66262 | .74896 | .67559 | .73728 | .68835 | .72537 | .70091 | .71325 | 30 |
| 31 | .64967 | .76022 | .66284 | .74876 | .67580 | .73708 | .68857 | .72517 | .70112 | .71305 | 29 |
| 32 | .64989 | .76003 | .66306 | .74857 | .67602 | .73688 | .68878 | .72497 | .70132 | .71284 | 28 |
| 33 | .65011 | .75984 | .66327 | .74838 | .67623 | .73669 | .68899 | .72477 | .70153 | .71264 | 27 |
| 34 | .65033 | .75965 | .66349 | .74818 | .67645 | .73649 | .68920 | .72457 | .70174 | .71243 | 26 |
| 35 | .65055 | .75946 | .66371 | .74799 | .67666 | .73629 | .68941 | .72437 | .70195 | .71223 | 25 |
| 36 | .65077 | .75927 | .66393 | .74780 | .67688 | .73610 | .68962 | .72417 | .70215 | .71203 | 24 |
| 37 | .65100 | .75908 | .66414 | .74760 | .67709 | .73590 | .68983 | .72397 | .70236 | .71182 | 23 |
| 38 | .65122 | .75889 | .66436 | .74741 | .67730 | .73570 | .69004 | .72377 | .70257 | .71162 | 22 |
| 39 | .65144 | .75870 | .66458 | .74722 | .67752 | .73551 | .69025 | .72357 | .70277 | .71141 | 21 |
| 40 | .65166 | .75851 | .66480 | .74703 | .67773 | .73531 | .69046 | .72337 | .70298 | .71121 | 20 |
| 41 | .65188 | .75832 | .66501 | .74683 | .67795 | .73511 | .69067 | .72317 | .70319 | .71100 | 19 |
| 42 | .65210 | .75813 | .66523 | .74664 | .67816 | .73491 | .69088 | .72297 | .70339 | .71080 | 18 |
| 43 | .65232 | .75794 | .66545 | .74644 | .67837 | .73472 | .69109 | .72277 | .70360 | .71059 | 17 |
| 44 | .65254 | .75775 | .66566 | .74625 | .67859 | .73452 | .69130 | .72257 | .70381 | .71039 | 16 |
| 45 | .65276 | .75756 | .66588 | .74606 | .67880 | .73432 | .69151 | .72236 | .70401 | .71019 | 15 |
| 46 | .65298 | .75738 | .66610 | .74586 | .67901 | .73413 | .69172 | .72216 | .70422 | .70998 | 14 |
| 47 | .65320 | .75719 | .66632 | .74567 | .67923 | .73393 | .69193 | .72196 | .70443 | .70978 | 13 |
| 48 | .65342 | .75700 | .66653 | .74548 | .67944 | .73373 | .69214 | .72176 | .70463 | .70957 | 12 |
| 49 | .65364 | .75680 | .66675 | .74528 | .67965 | .73353 | .69235 | .72156 | .70484 | .70937 | 11 |
| 50 | .65386 | .75661 | .66697 | .74509 | .67987 | .73333 | .69256 | .72136 | .70505 | .70916 | 10 |
| 51 | .65408 | .75642 | .66718 | .74489 | .68008 | .73314 | .69277 | .72116 | .70525 | .70896 | 9 |
| 52 | .65430 | .75623 | .66740 | .74470 | .68029 | .73294 | .69298 | .72095 | .70546 | .70875 | 8 |
| 53 | .65452 | .75604 | .66762 | .74451 | .68051 | .73274 | .69319 | .72075 | .70567 | .70855 | 7 |
| 54 | .65474 | .75585 | .66783 | .74431 | .68072 | .73254 | .69340 | .72055 | .70587 | .70834 | 6 |
| 55 | .65496 | .75566 | .66805 | .74412 | .68093 | .73234 | .69361 | .72035 | .70608 | .70813 | 5 |
| 56 | .65518 | .75547 | .66827 | .74392 | .68115 | .73215 | .69382 | .72015 | .70628 | .70793 | 4 |
| 57 | .65540 | .75528 | .66848 | .74373 | .68136 | .73195 | .69403 | .71995 | .70649 | .70772 | 3 |
| 58 | .65562 | .75509 | .66870 | .74353 | .68157 | .73175 | .69424 | .71974 | .70670 | .70752 | 2 |
| 59 | .65584 | .75490 | .66891 | .74334 | .68179 | .73155 | .69445 | .71954 | .70690 | .70731 | 1 |
| 60 | .65606 | .75471 | .66913 | .74314 | .68200 | .73135 | .69466 | .71934 | .70711 | .70711 | 0 |
| ′ | Cosin | Sine | Cosin | Sine | Cosin | Sine | Cosin | Sine | Cosin | Sine | ′ |
|   | 49° | | 48° | | 47° | | 46° | | 45° | | |

## NATURAL TANGENTS AND COTANGENTS. 121

| ′ | 0° | | 1° | | 2° | | 3° | | ′ |
|---|---|---|---|---|---|---|---|---|---|
|  | Tang | Cotang | Tang | Cotang | Tang | Cotang | Tang | Cotang |  |
| 0 | .00000 | Infinite. | .01746 | 57.2900 | .03492 | 28.6363 | .05241 | 19.0811 | 60 |
| 1 | .00029 | 3437.75 | .01775 | 56.3506 | .03521 | 28.3994 | .05270 | 18.9755 | 59 |
| 2 | .00058 | 1718.87 | .01804 | 55.4415 | .03550 | 28.1664 | .05299 | 18.8711 | 58 |
| 3 | .00087 | 1145.92 | .01833 | 54.5613 | .03579 | 27.9372 | .05328 | 18.7678 | 57 |
| 4 | .00116 | 859.436 | .01862 | 53.7086 | .03609 | 27.7117 | .05357 | 18.6656 | 56 |
| 5 | .00145 | 687.549 | .01891 | 52.8821 | .03638 | 27.4899 | .05387 | 18.5645 | 55 |
| 6 | .00175 | 572.957 | .01920 | 52.0807 | .03667 | 27.2715 | .05416 | 18.4645 | 54 |
| 7 | .00204 | 491.106 | .01949 | 51.3032 | .03696 | 27.0566 | .05445 | 18.3655 | 53 |
| 8 | .00233 | 429.718 | .01978 | 50.5485 | .03725 | 26.8450 | .05474 | 18.2677 | 52 |
| 9 | .00262 | 381.971 | .02007 | 49.8157 | .03754 | 26.6367 | .05503 | 18.1708 | 51 |
| 10 | .00291 | 343.774 | .02036 | 49.1039 | .03783 | 26.4316 | .05533 | 18.0750 | 50 |
| 11 | .00320 | 312.521 | .02066 | 48.4121 | .03812 | 26.2296 | .05562 | 17.9802 | 49 |
| 12 | .00349 | 286.478 | .02095 | 47.7395 | .03842 | 26.0307 | .05591 | 17.8863 | 48 |
| 13 | .00378 | 264.441 | .02124 | 47.0853 | .03871 | 25.8348 | .05620 | 17.7934 | 47 |
| 14 | .00407 | 245.552 | .02153 | 46.4489 | .03900 | 25.6418 | .05649 | 17.7015 | 46 |
| 15 | .00436 | 229.182 | .02182 | 45.8294 | .03929 | 25.4517 | .05678 | 17.6106 | 45 |
| 16 | .00465 | 214.858 | .02211 | 45.2261 | .03958 | 25.2644 | .05708 | 17.5205 | 44 |
| 17 | .00495 | 202.219 | .02240 | 44.6386 | .03987 | 25.0798 | .05737 | 17.4314 | 43 |
| 18 | .00524 | 190.984 | .02269 | 44.0661 | .04016 | 24.8978 | .05766 | 17.3432 | 42 |
| 19 | .00553 | 180.932 | .02298 | 43.5081 | .04046 | 24.7185 | .05795 | 17.2558 | 41 |
| 20 | .00582 | 171.885 | .02328 | 42.9641 | .04075 | 24.5418 | .05824 | 17.1693 | 40 |
| 21 | .00611 | 163.700 | .02357 | 42.4335 | .04104 | 24.3675 | .05854 | 17.0837 | 39 |
| 22 | .00640 | 156.259 | .02386 | 41.9158 | .04133 | 24.1957 | .05883 | 16.9990 | 38 |
| 23 | .00669 | 149.465 | .02415 | 41.4106 | .04162 | 24.0263 | .05912 | 16.9150 | 37 |
| 24 | .00698 | 143.237 | .02444 | 40.9174 | .04191 | 23.8593 | .05941 | 16.8319 | 36 |
| 25 | .00727 | 137.507 | .02473 | 40.4358 | .04220 | 23.6945 | .05970 | 16.7496 | 35 |
| 26 | .00756 | 132.219 | .02502 | 39.9655 | .04250 | 23.5321 | .05999 | 16.6681 | 34 |
| 27 | .00785 | 127.321 | .02531 | 39.5059 | .04279 | 23.3718 | .06029 | 16.5874 | 33 |
| 28 | .00815 | 122.774 | .02560 | 39.0568 | .04308 | 23.2137 | .06058 | 16.5075 | 32 |
| 29 | .00844 | 118.540 | .02589 | 38.6177 | .04337 | 23.0577 | .06087 | 16.4283 | 31 |
| 30 | .00873 | 114.589 | .02619 | 38.1885 | .04366 | 22.9038 | .06116 | 16.3499 | 30 |
| 31 | .00902 | 110.892 | .02648 | 37.7686 | .04395 | 22.7519 | .06145 | 16.2722 | 29 |
| 32 | .00931 | 107.426 | .02677 | 37.3579 | .04424 | 22.6020 | .06175 | 16.1952 | 28 |
| 33 | .00960 | 104.171 | .02706 | 36.9560 | .04454 | 22.4541 | .06204 | 16.1190 | 27 |
| 34 | .00989 | 101.107 | .02735 | 36.5627 | .04483 | 22.3081 | .06233 | 16.0435 | 26 |
| 35 | .01018 | 98.2179 | .02764 | 36.1776 | .04512 | 22.1640 | .06262 | 15.9687 | 25 |
| 36 | .01047 | 95.4895 | .02793 | 35.8006 | .04541 | 22.0217 | .06291 | 15.8945 | 24 |
| 37 | .01076 | 92.9085 | .02822 | 35.4313 | .04570 | 21.8813 | .06321 | 15.8211 | 23 |
| 38 | .01105 | 90.4633 | .02851 | 35.0695 | .04599 | 21.7426 | .06350 | 15.7483 | 22 |
| 39 | .01135 | 88.1436 | .02881 | 34.7151 | .04628 | 21.6056 | .06379 | 15.6762 | 21 |
| 40 | .01164 | 85.9398 | .02910 | 34.3678 | .04658 | 21.4704 | .06408 | 15.6048 | 20 |
| 41 | .01193 | 83.8435 | .02939 | 34.0273 | .04687 | 21.3369 | .06437 | 15.5340 | 19 |
| 42 | .01222 | 81.8470 | .02968 | 33.6935 | .04716 | 21.2049 | .06467 | 15.4638 | 18 |
| 43 | .01251 | 79.9434 | .02997 | 33.3662 | .04745 | 21.0747 | .06496 | 15.3943 | 17 |
| 44 | .01280 | 78.1263 | .03026 | 33.0452 | .04774 | 20.9460 | .06525 | 15.3254 | 16 |
| 45 | .01309 | 76.3900 | .03055 | 32.7303 | .04803 | 20.8188 | .06554 | 15.2571 | 15 |
| 46 | .01338 | 74.7292 | .03084 | 32.4213 | .04833 | 20.6932 | .06584 | 15.1893 | 14 |
| 47 | .01367 | 73.1390 | .03114 | 32.1181 | .04862 | 20.5691 | .06613 | 15.1222 | 13 |
| 48 | .01396 | 71.6151 | .03143 | 31.8205 | .04891 | 20.4465 | .06642 | 15.0557 | 12 |
| 49 | .01425 | 70.1533 | .03172 | 31.5284 | .04920 | 20.3253 | .06671 | 14.9898 | 11 |
| 50 | .01455 | 68.7501 | .03201 | 31.2416 | .04949 | 20.2056 | .06700 | 14.9244 | 10 |
| 51 | .01484 | 67.4019 | .03230 | 30.9599 | .04978 | 20.0872 | .06730 | 14.8596 | 9 |
| 52 | .01513 | 66.1055 | .03259 | 30.6833 | .05007 | 19.9702 | .06759 | 14.7954 | 8 |
| 53 | .01542 | 64.8580 | .03288 | 30.4116 | .05037 | 19.8546 | .06788 | 14.7317 | 7 |
| 54 | .01571 | 63.6567 | .03317 | 30.1446 | .05066 | 19.7403 | .06817 | 14.6685 | 6 |
| 55 | .01600 | 62.4992 | .03346 | 29.8823 | .05095 | 19.6273 | .06847 | 14.6059 | 5 |
| 56 | .01629 | 61.3829 | .03376 | 29.6245 | .05124 | 19.5156 | .06876 | 14.5438 | 4 |
| 57 | .01658 | 60.3058 | .03405 | 29.3711 | .05153 | 19.4051 | .06905 | 14.4823 | 3 |
| 58 | .01687 | 59.2659 | .03434 | 29.1220 | .05182 | 19.2959 | .06934 | 14.4212 | 2 |
| 59 | .01716 | 58.2612 | .03463 | 28.8771 | .05212 | 19.1879 | .06963 | 14.3607 | 1 |
| 60 | .01746 | 57.2900 | .03492 | 28.6363 | .05241 | 19.0811 | .06993 | 14.3007 | 0 |
|  | Cotang | Tang | Cotang | Tang | Cotang | Tang | Cotang | Tang |  |
| ′ | 89° | | 88° | | 87° | | 86° | | ′ |

## NATURAL TANGENTS AND COTANGENTS.

| ′ | 4° Tang | 4° Cotang | 5° Tang | 5° Cotang | 6° Tang | 6° Cotang | 7° Tang | 7° Cotang | ′ |
|---|---|---|---|---|---|---|---|---|---|
| 0 | .06993 | 14.3007 | .08749 | 11.4301 | .10510 | 9.51436 | .12278 | 8.14435 | 60 |
| 1 | .07022 | 14.2411 | .08778 | 11.3919 | .10540 | 9.48781 | .12308 | 8.12481 | 59 |
| 2 | .07051 | 14.1821 | .08807 | 11.3540 | .10569 | 9.46141 | .12338 | 8.10536 | 58 |
| 3 | .07080 | 14.1235 | .08837 | 11.3163 | .10599 | 9.43515 | .12367 | 8.08600 | 57 |
| 4 | .07110 | 14.0655 | .08866 | 11.2789 | .10628 | 9.40904 | .12397 | 8.06674 | 56 |
| 5 | .07139 | 14.0079 | .08895 | 11.2417 | .10657 | 9.38307 | .12426 | 8.04756 | 55 |
| 6 | .07168 | 13.9507 | .08925 | 11.2048 | .10687 | 9.35724 | .12456 | 8.02848 | 54 |
| 7 | .07197 | 13.8940 | .08954 | 11.1681 | .10716 | 9.33155 | .12485 | 8.00948 | 53 |
| 8 | .07227 | 13.8378 | .08983 | 11.1316 | .10746 | 9.30599 | .12515 | 7.99058 | 52 |
| 9 | .07256 | 13.7821 | .09013 | 11.0954 | .10775 | 9.28058 | .12544 | 7.97176 | 51 |
| 10 | .07285 | 13.7267 | .09042 | 11.0594 | .10805 | 9.25530 | .12574 | 7.95302 | 50 |
| 11 | .07314 | 13.6719 | .09071 | 11.0237 | .10834 | 9.23016 | .12603 | 7.93438 | 49 |
| 12 | .07344 | 13.6174 | .09101 | 10.9882 | .10863 | 9.20516 | .12633 | 7.91582 | 48 |
| 13 | .07373 | 13.5634 | .09130 | 10.9529 | .10893 | 9.18028 | .12662 | 7.89734 | 47 |
| 14 | .07402 | 13.5098 | .09159 | 10.9178 | .10922 | 9.15554 | .12692 | 7.87895 | 46 |
| 15 | .07431 | 13.4566 | .09189 | 10.8829 | .10952 | 9.13093 | .12722 | 7.86064 | 45 |
| 16 | .07461 | 13.4039 | .09218 | 10.8483 | .10981 | 9.10646 | .12751 | 7.84242 | 44 |
| 17 | .07490 | 13.3515 | .09247 | 10.8139 | .11011 | 9.08211 | .12781 | 7.82428 | 43 |
| 18 | .07519 | 13.2996 | .09277 | 10.7797 | .11040 | 9.05789 | .12810 | 7.80622 | 42 |
| 19 | .07548 | 13.2480 | .09306 | 10.7457 | .11070 | 9.03379 | .12840 | 7.78825 | 41 |
| 20 | .07578 | 13.1969 | .09335 | 10.7119 | .11099 | 9.00983 | .12869 | 7.77035 | 40 |
| 21 | .07607 | 13.1461 | .09365 | 10.6783 | .11128 | 8.98598 | .12899 | 7.75254 | 39 |
| 22 | .07636 | 13.0958 | .09394 | 10.6450 | .11158 | 8.96227 | .12929 | 7.73480 | 38 |
| 23 | .07665 | 13.0458 | .09423 | 10.6118 | .11187 | 8.93867 | .12958 | 7.71715 | 37 |
| 24 | .07695 | 12.9962 | .09453 | 10.5789 | .11217 | 8.91520 | .12988 | 7.69957 | 36 |
| 25 | .07724 | 12.9469 | .09482 | 10.5462 | .11246 | 8.89185 | .13017 | 7.68208 | 35 |
| 26 | .07753 | 12.8981 | .09511 | 10.5136 | .11276 | 8.86862 | .13047 | 7.66466 | 34 |
| 27 | .07782 | 12.8496 | .09541 | 10.4813 | .11305 | 8.84551 | .13076 | 7.64732 | 33 |
| 28 | .07812 | 12.8014 | .09570 | 10.4491 | .11335 | 8.82252 | .13106 | 7.63005 | 32 |
| 29 | .07841 | 12.7536 | .09600 | 10.4172 | .11364 | 8.79964 | .13136 | 7.61287 | 31 |
| 30 | .07870 | 12.7062 | .09629 | 10.3854 | .11394 | 8.77689 | .13165 | 7.59575 | 30 |
| 31 | .07899 | 12.6591 | .09658 | 10.3538 | .11423 | 8.75425 | .13195 | 7.57872 | 29 |
| 32 | .07929 | 12.6124 | .09688 | 10.3224 | .11452 | 8.73172 | .13224 | 7.56176 | 28 |
| 33 | .07958 | 12.5660 | .09717 | 10.2913 | .11482 | 8.70931 | .13254 | 7.54487 | 27 |
| 34 | .07987 | 12.5199 | .09746 | 10.2602 | .11511 | 8.68701 | .13284 | 7.52806 | 26 |
| 35 | .08017 | 12.4742 | .09776 | 10.2294 | .11541 | 8.66482 | .13313 | 7.51132 | 25 |
| 36 | .08046 | 12.4288 | .09805 | 10.1988 | .11570 | 8.64275 | .13343 | 7.49465 | 24 |
| 37 | .08075 | 12.3838 | .09834 | 10.1683 | .11600 | 8.62078 | .13372 | 7.47806 | 23 |
| 38 | .08104 | 12.3390 | .09864 | 10.1381 | .11629 | 8.59893 | .13402 | 7.46154 | 22 |
| 39 | .08134 | 12.2946 | .09893 | 10.1080 | .11659 | 8.57718 | .13432 | 7.44509 | 21 |
| 40 | .08163 | 12.2505 | .09923 | 10.0780 | .11688 | 8.55555 | .13461 | 7.42871 | 20 |
| 41 | .08192 | 12.2067 | .09952 | 10.0483 | .11718 | 8.53402 | .13491 | 7.41240 | 19 |
| 42 | .08221 | 12.1632 | .09981 | 10.0187 | .11747 | 8.51259 | .13521 | 7.39616 | 18 |
| 43 | .08251 | 12.1201 | .10011 | 9.98931 | .11777 | 8.49128 | .13550 | 7.37999 | 17 |
| 44 | .08280 | 12.0772 | .10040 | 9.96007 | .11806 | 8.47007 | .13580 | 7.36389 | 16 |
| 45 | .08309 | 12.0346 | .10069 | 9.93101 | .11836 | 8.44896 | .13609 | 7.34786 | 15 |
| 46 | .08339 | 11.9923 | .10099 | 9.90211 | .11865 | 8.42795 | .13639 | 7.33190 | 14 |
| 47 | .08368 | 11.9504 | .10128 | 9.87338 | .11895 | 8.40705 | .13669 | 7.31600 | 13 |
| 48 | .08397 | 11.9087 | .10158 | 9.84482 | .11924 | 8.38625 | .13698 | 7.30018 | 12 |
| 49 | .08427 | 11.8673 | .10187 | 9.81641 | .11954 | 8.36555 | .13728 | 7.28442 | 11 |
| 50 | .08456 | 11.8262 | .10216 | 9.78817 | .11983 | 8.34496 | .13758 | 7.26873 | 10 |
| 51 | .08485 | 11.7853 | .10246 | 9.76009 | .12013 | 8.32446 | .13787 | 7.25310 | 9 |
| 52 | .08514 | 11.7448 | .10275 | 9.73217 | .12042 | 8.30406 | .13817 | 7.23754 | 8 |
| 53 | .08544 | 11.7045 | .10305 | 9.70441 | .12072 | 8.28376 | .13846 | 7.22204 | 7 |
| 54 | .08573 | 11.6645 | .10334 | 9.67680 | .12101 | 8.26355 | .13876 | 7.20661 | 6 |
| 55 | .08602 | 11.6248 | .10363 | 9.64935 | .12131 | 8.24345 | .13906 | 7.19125 | 5 |
| 56 | .08632 | 11.5853 | .10393 | 9.62205 | .12160 | 8.22344 | .13935 | 7.17594 | 4 |
| 57 | .08661 | 11.5461 | .10422 | 9.59490 | .12190 | 8.20352 | .13965 | 7.16071 | 3 |
| 58 | .08690 | 11.5072 | .10452 | 9.56791 | .12219 | 8.18370 | .13995 | 7.14553 | 2 |
| 59 | .08720 | 11.4685 | .10481 | 9.54106 | .12249 | 8.16398 | .14024 | 7.13042 | 1 |
| 60 | .08749 | 11.4301 | .10510 | 9.51436 | .12278 | 8.14435 | .14054 | 7.11537 | 0 |
| ′ | Cotang | Tang | Cotang | Tang | Cotang | Tang | Cotang | Tang | ′ |
|  | 85° | | 84° | | 83° | | 82° | | |

## NATURAL TANGENTS AND COTANGENTS.

| ′ | 8° Tang | 8° Cotang | 9° Tang | 9° Cotang | 10° Tang | 10° Cotang | 11° Tang | 11° Cotang | ′ |
|---|---|---|---|---|---|---|---|---|---|
| 0 | .14054 | 7.11537 | .15838 | 6.31375 | .17633 | 5.67128 | .19438 | 5.14455 | 60 |
| 1 | .14084 | 7.10038 | .15868 | 6.30189 | .17663 | 5.66165 | .19468 | 5.13658 | 59 |
| 2 | .14113 | 7.08546 | .15898 | 6.29007 | .17693 | 5.65205 | .19498 | 5.12862 | 58 |
| 3 | .14143 | 7.07059 | .15928 | 6.27829 | .17723 | 5.64248 | .19529 | 5.12069 | 57 |
| 4 | .14173 | 7.05579 | .15958 | 6.26655 | .17753 | 5.63295 | .19559 | 5.11279 | 56 |
| 5 | .14202 | 7.04105 | .15988 | 6.25486 | .17783 | 5.62344 | .19589 | 5.10490 | 55 |
| 6 | .14232 | 7.02637 | .16017 | 6.24321 | .17813 | 5.61397 | .19619 | 5.09704 | 54 |
| 7 | .14262 | 6.91174 | .16047 | 6.23160 | .17843 | 5.60452 | .19649 | 5.08921 | 53 |
| 8 | .14291 | 6.99718 | .16077 | 6.22003 | .17873 | 5.59511 | .19680 | 5.08139 | 52 |
| 9 | .14321 | 6.98268 | .16107 | 6.20851 | .17903 | 5.58573 | .19710 | 5.07360 | 51 |
| 10 | .14351 | 6.96823 | .16137 | 6.19703 | .17933 | 5.57638 | .19740 | 5.06584 | 50 |
| 11 | .14381 | 6.95385 | .16167 | 6.18559 | .17963 | 5.56706 | .19770 | 5.05809 | 49 |
| 12 | .14410 | 6.93952 | .16196 | 6.17419 | .17993 | 5.55777 | .19801 | 5.05037 | 48 |
| 13 | .14440 | 6.92525 | .16226 | 6.16283 | .18023 | 5.54851 | .19831 | 5.04267 | 47 |
| 14 | .14470 | 6.91104 | .16256 | 6.15151 | .18053 | 5.53927 | .19861 | 5.03499 | 46 |
| 15 | .14499 | 6.89688 | .16286 | 6.14023 | .18083 | 5.53007 | .19891 | 5.02734 | 45 |
| 16 | .14529 | 6.88278 | .16316 | 6.12899 | .18113 | 5.52090 | .19921 | 5.01971 | 44 |
| 17 | .14559 | 6.86874 | .16346 | 6.11779 | .18143 | 5.51176 | .19952 | 5.01210 | 43 |
| 18 | .14588 | 6.85475 | .16376 | 6.10664 | .18173 | 5.50264 | .19982 | 5.00451 | 42 |
| 19 | .14618 | 6.84082 | .16405 | 6.09552 | .18203 | 5.49356 | .20012 | 4.99695 | 41 |
| 20 | .14648 | 6.82694 | .16435 | 6.08444 | .18233 | 5.48451 | .20042 | 4.98940 | 40 |
| 21 | .14678 | 6.81312 | .16465 | 6.07340 | .18263 | 5.47548 | .20073 | 4.98188 | 39 |
| 22 | .14707 | 6.79936 | .16495 | 6.06240 | .18293 | 5.46648 | .20103 | 4.97438 | 38 |
| 23 | .14737 | 6.78564 | .16525 | 6.05143 | .18323 | 5.45751 | .20133 | 4.96690 | 37 |
| 24 | .14767 | 6.77199 | .16555 | 6.04051 | .18353 | 5.44857 | .20164 | 4.95945 | 36 |
| 25 | .14796 | 6.75838 | .16585 | 6.02962 | .18384 | 5.43966 | .20194 | 4.95201 | 35 |
| 26 | .14826 | 6.74483 | .16615 | 6.01878 | .18414 | 5.43077 | .20224 | 4.94460 | 34 |
| 27 | .14856 | 6.73133 | .16645 | 6.00797 | .18444 | 5.42192 | .20254 | 4.93721 | 33 |
| 28 | .14886 | 6.71789 | .16674 | 5.99720 | .18474 | 5.41309 | .20285 | 4.92984 | 32 |
| 29 | .14915 | 6.70450 | .16704 | 5.98646 | .18504 | 5.40429 | .20315 | 4.92249 | 31 |
| 30 | .14945 | 6.69116 | .16734 | 5.97576 | .18534 | 5.39552 | .20345 | 4.91516 | 30 |
| 31 | .14975 | 6.67787 | .16764 | 5.96510 | .18564 | 5.38677 | .20376 | 4.90785 | 29 |
| 32 | .15005 | 6.66463 | .16794 | 5.95448 | .18594 | 5.37805 | .20406 | 4.90056 | 28 |
| 33 | .15034 | 6.65144 | .16824 | 5.94390 | .18624 | 5.36936 | .20436 | 4.89330 | 27 |
| 34 | .15064 | 6.63831 | .16854 | 5.93335 | .18654 | 5.36070 | .20466 | 4.88605 | 26 |
| 35 | .15094 | 6.62523 | .16884 | 5.92283 | .18684 | 5.35206 | .20497 | 4.87882 | 25 |
| 36 | .15124 | 6.61219 | .16914 | 5.91236 | .18714 | 5.34345 | .20527 | 4.87162 | 24 |
| 37 | .15153 | 6.59921 | .16944 | 5.90191 | .18745 | 5.33487 | .20557 | 4.86444 | 23 |
| 38 | .15183 | 6.58627 | .16974 | 5.89151 | .18775 | 5.32631 | .20588 | 4.85727 | 22 |
| 39 | .15213 | 6.57339 | .17004 | 5.88114 | .18805 | 5.31778 | .20618 | 4.85013 | 21 |
| 40 | .15243 | 6.56055 | .17033 | 5.87080 | .18835 | 5.30928 | .20648 | 4.84300 | 20 |
| 41 | .15272 | 6.54777 | .17063 | 5.86051 | .18865 | 5.30080 | .20679 | 4.83590 | 19 |
| 42 | .15302 | 6.53503 | .17093 | 5.85024 | .18895 | 5.29235 | .20709 | 4.82882 | 18 |
| 43 | .15332 | 6.52234 | .17123 | 5.84001 | .18925 | 5.28393 | .20739 | 4.82175 | 17 |
| 44 | .15362 | 6.50970 | .17153 | 5.82982 | .18955 | 5.27553 | .20770 | 4.81471 | 16 |
| 45 | .15391 | 6.49710 | .17183 | 5.81966 | .18986 | 5.26715 | .20800 | 4.80769 | 15 |
| 46 | .15421 | 6.48456 | .17213 | 5.80953 | .19016 | 5.25880 | .20830 | 4.80068 | 14 |
| 47 | .15451 | 6.47206 | .17243 | 5.79944 | .19046 | 5.25048 | .20861 | 4.79370 | 13 |
| 48 | .15481 | 6.45961 | .17273 | 5.78938 | .19076 | 5.24218 | .20891 | 4.78673 | 12 |
| 49 | .15511 | 6.44720 | .17303 | 5.77936 | .19106 | 5.23391 | .20921 | 4.77978 | 11 |
| 50 | .15540 | 6.43484 | .17333 | 5.76937 | .19136 | 5.22566 | .20952 | 4.77286 | 10 |
| 51 | .15570 | 6.42253 | .17363 | 5.75941 | .19166 | 5.21744 | .20982 | 4.76595 | 9 |
| 52 | .15600 | 6.41026 | .17393 | 5.74949 | .19197 | 5.20925 | .21013 | 4.75906 | 8 |
| 53 | .15630 | 6.39804 | .17423 | 5.73960 | .19227 | 5.20107 | .21043 | 4.75219 | 7 |
| 54 | .15660 | 6.38587 | .17453 | 5.72974 | .19257 | 5.19293 | .21073 | 4.74534 | 6 |
| 55 | .15689 | 6.37374 | .17483 | 5.71992 | .19287 | 5.18480 | .21104 | 4.73851 | 5 |
| 56 | .15719 | 6.36165 | .17513 | 5.71013 | .19317 | 5.17671 | .21134 | 4.73170 | 4 |
| 57 | .15749 | 6.34961 | .17543 | 5.70037 | .19347 | 5.16863 | .21164 | 4.72490 | 3 |
| 58 | .15779 | 6.33761 | .17573 | 5.69064 | .19378 | 5.16058 | .21195 | 4.71813 | 2 |
| 59 | .15809 | 6.32566 | .17603 | 5.68094 | .19408 | 5.15256 | .21225 | 4.71137 | 1 |
| 60 | .15838 | 6.31375 | .17633 | 5.67128 | .19438 | 5.14455 | .21256 | 4.70463 | 0 |
| | Cotang | Tang | Cotang | Tang | Cotang | Tang | Cotang | Tang | ′ |
| | 81° | | 80° | | 79° | | 78° | | |

## NATURAL TANGENTS AND COTANGENTS.

| ′ | 12° | | 13° | | 14° | | 15° | | ′ |
|---|---|---|---|---|---|---|---|---|---|
| | Tang | Cotang | Tang | Cotang | Tang | Cotang | Tang | Cotang | |
| 0 | .21256 | 4.70463 | .23087 | 4.33148 | .24933 | 4.01078 | .26795 | 3.73205 | 60 |
| 1 | .21286 | 4.69791 | .23117 | 4.32573 | .24964 | 4.00582 | .26826 | 3.72771 | 59 |
| 2 | .21316 | 4.69121 | .23148 | 4.32001 | .24995 | 4.00086 | .26857 | 3.72338 | 58 |
| 3 | .21347 | 4.68452 | .23179 | 4.31430 | .25026 | 3.99592 | .26888 | 3.71907 | 57 |
| 4 | .21377 | 4.67786 | .23209 | 4.30860 | .25056 | 3.99099 | .26920 | 3.71476 | 56 |
| 5 | .21408 | 4.67121 | .23240 | 4.30291 | .25087 | 3.98607 | .26951 | 3.71046 | 55 |
| 6 | .21438 | 4.66458 | .23271 | 4.29724 | .25118 | 3.98117 | .26982 | 3.70616 | 54 |
| 7 | .21469 | 4.65797 | .23301 | 4.29159 | .25149 | 3.97627 | .27013 | 3.70188 | 53 |
| 8 | .21499 | 4.65138 | .23332 | 4.28595 | .25180 | 3.97139 | .27044 | 3.69761 | 52 |
| 9 | .21529 | 4.64480 | .23363 | 4.28032 | .25211 | 3.96651 | .27076 | 3.69335 | 51 |
| 10 | .21560 | 4.63825 | .23393 | 4.27471 | .25242 | 3.96165 | .27107 | 3.68909 | 50 |
| 11 | .21590 | 4.63171 | .23424 | 4.26911 | .25273 | 3.95680 | .27138 | 3.68485 | 49 |
| 12 | .21621 | 4.62518 | .23455 | 4.26352 | .25304 | 3.95196 | .27169 | 3.68061 | 48 |
| 13 | .21651 | 4.61868 | .23485 | 4.25795 | .25335 | 3.94713 | .27201 | 3.67638 | 47 |
| 14 | .21682 | 4.61219 | .23516 | 4.25239 | .25366 | 3.94232 | .27232 | 3.67217 | 46 |
| 15 | .21712 | 4.60572 | .23547 | 4.24685 | .25397 | 3.93751 | .27263 | 3.66796 | 45 |
| 16 | .21743 | 4.59927 | .23578 | 4.24132 | .25428 | 3.93271 | .27294 | 3.66376 | 44 |
| 17 | .21773 | 4.59283 | .23608 | 4.23580 | .25459 | 3.92793 | .27326 | 3.65957 | 43 |
| 18 | .21804 | 4.58641 | .23639 | 4.23030 | .25490 | 3.92316 | .27357 | 3.65538 | 42 |
| 19 | .21834 | 4.58001 | .23670 | 4.22481 | .25521 | 3.91839 | .27388 | 3.65121 | 41 |
| 20 | .21864 | 4.57363 | .23700 | 4.21933 | .25552 | 3.91364 | .27419 | 3.64705 | 40 |
| 21 | .21895 | 4.56726 | .23731 | 4.21387 | .25583 | 3.90890 | .27451 | 3.64289 | 39 |
| 22 | .21925 | 4.56091 | .23762 | 4.20842 | .25614 | 3.90417 | .27482 | 3.63874 | 38 |
| 23 | .21956 | 4.55458 | .23793 | 4.20298 | .25645 | 3.89945 | .27513 | 3.63461 | 37 |
| 24 | .21986 | 4.54826 | .23823 | 4.19756 | .25676 | 3.89474 | .27545 | 3.63048 | 36 |
| 25 | .22017 | 4.54196 | .23854 | 4.19215 | .25707 | 3.89004 | .27576 | 3.62636 | 35 |
| 26 | .22047 | 4.53568 | .23885 | 4.18675 | .25738 | 3.88536 | .27607 | 3.62224 | 34 |
| 27 | .22078 | 4.52941 | .23916 | 4.18137 | .25769 | 3.88068 | .27638 | 3.61814 | 33 |
| 28 | .22108 | 4.52316 | .23946 | 4.17600 | .25800 | 3.87601 | .27670 | 3.61405 | 32 |
| 29 | .22139 | 4.51693 | .23977 | 4.17064 | .25831 | 3.87136 | .27701 | 3.60996 | 31 |
| 30 | .22169 | 4.51071 | .24008 | 4.16530 | .25862 | 3.86671 | .27732 | 3.60588 | 30 |
| 31 | .22200 | 4.50451 | .24039 | 4.15997 | .25893 | 3.86208 | .27764 | 3.60181 | 29 |
| 32 | .22231 | 4.49832 | .24069 | 4.15465 | .25924 | 3.85745 | .27795 | 3.59775 | 28 |
| 33 | .22261 | 4.49215 | .24100 | 4.14934 | .25955 | 3.85284 | .27826 | 3.59370 | 27 |
| 34 | .22292 | 4.48600 | .24131 | 4.14405 | .25986 | 3.84824 | .27858 | 3.58966 | 26 |
| 35 | .22322 | 4.47986 | .24162 | 4.13877 | .26017 | 3.84364 | .27889 | 3.58562 | 25 |
| 36 | .22353 | 4.47374 | .24193 | 4.13350 | .26048 | 3.83906 | .27921 | 3.58160 | 24 |
| 37 | .22383 | 4.46764 | .24223 | 4.12825 | .26079 | 3.83449 | .27952 | 3.57758 | 23 |
| 38 | .22414 | 4.46155 | .24254 | 4.12301 | .26110 | 3.82992 | .27983 | 3.57357 | 22 |
| 39 | .22444 | 4.45548 | .24285 | 4.11778 | .26141 | 3.82537 | .28015 | 3.56957 | 21 |
| 40 | .22475 | 4.44942 | .24316 | 4.11256 | .26172 | 3.82083 | .28046 | 3.56557 | 20 |
| 41 | .22505 | 4.44338 | .24347 | 4.10736 | .26203 | 3.81630 | .28077 | 3.56159 | 19 |
| 42 | .22536 | 4.43735 | .24377 | 4.10216 | .26235 | 3.81177 | .28109 | 3.55761 | 18 |
| 43 | .22567 | 4.43134 | .24408 | 4.09699 | .26266 | 3.80726 | .28140 | 3.55364 | 17 |
| 44 | .22597 | 4.42534 | .24439 | 4.09182 | .26297 | 3.80276 | .28172 | 3.54968 | 16 |
| 45 | .22628 | 4.41936 | .24470 | 4.08666 | .26328 | 3.79827 | .28203 | 3.54573 | 15 |
| 46 | .22658 | 4.41340 | .24501 | 4.08152 | .26359 | 3.79378 | .28234 | 3.54179 | 14 |
| 47 | .22689 | 4.40745 | .24532 | 4.07639 | .26390 | 3.78931 | .28266 | 3.53785 | 13 |
| 48 | .22719 | 4.40152 | .24562 | 4.07127 | .26421 | 3.78485 | .28297 | 3.53393 | 12 |
| 49 | .22750 | 4.39560 | .24593 | 4.06616 | .26452 | 3.78040 | .28329 | 3.53001 | 11 |
| 50 | .22781 | 4.38969 | .24624 | 4.06107 | .26483 | 3.77595 | .28360 | 3.52609 | 10 |
| 51 | .22811 | 4.38381 | .24655 | 4.05599 | .26515 | 3.77152 | .28391 | 3.52219 | 9 |
| 52 | .22842 | 4.37793 | .24686 | 4.05092 | .26546 | 3.76709 | .28423 | 3.51829 | 8 |
| 53 | .22872 | 4.37207 | .24717 | 4.04586 | .26577 | 3.76268 | .28454 | 3.51441 | 7 |
| 54 | .22903 | 4.36623 | .24747 | 4.04081 | .26608 | 3.75828 | .28486 | 3.51053 | 6 |
| 55 | .22934 | 4.36040 | .24778 | 4.03578 | .26639 | 3.75388 | .28517 | 3.50666 | 5 |
| 56 | .22964 | 4.35459 | .24809 | 4.03076 | .26670 | 3.74950 | .28549 | 3.50279 | 4 |
| 57 | .22995 | 4.34879 | .24840 | 4.02574 | .26701 | 3.74512 | .28580 | 3.49894 | 3 |
| 58 | .23026 | 4.34300 | .24871 | 4.02074 | .26733 | 3.74075 | .28612 | 3.49509 | 2 |
| 59 | .23056 | 4.33723 | .24902 | 4.01576 | .26764 | 3.73640 | .28643 | 3.49125 | 1 |
| 60 | .23087 | 4.33148 | .24933 | 4.01078 | .26795 | 3.73205 | .28675 | 3.48741 | 0 |
| ′ | Cotang | Tang | Cotang | Tang | Cotang | Tang | Cotang | Tang | ′ |
| | 77° | | 76° | | 75° | | 74° | | |

## NATURAL TANGENTS AND COTANGENTS. 125

| ′ | 16° | | 17° | | 18° | | 19° | | ′ |
|---|---|---|---|---|---|---|---|---|---|
| | Tang | Cotang | Tang | Cotang | Tang | Cotang | Tang | Cotang | |
| 0 | .28675 | 3.48741 | .30573 | 3.27085 | .32492 | 3.07768 | .34433 | 2.90421 | 60 |
| 1 | .28706 | 3.48359 | .30605 | 3.26745 | .32524 | 3.07464 | .34465 | 2.90147 | 59 |
| 2 | .28738 | 3.47977 | .30637 | 3.26406 | .32556 | 3.07160 | .34498 | 2.89873 | 58 |
| 3 | .28769 | 3.47596 | .30669 | 3.26067 | .32588 | 3.06857 | .34530 | 2.89600 | 57 |
| 4 | .28800 | 3.47216 | .30700 | 3.25729 | .32621 | 3.06554 | .34563 | 2.89327 | 56 |
| 5 | .28832 | 3.46837 | .30732 | 3.25392 | .32653 | 3.06252 | .34596 | 2.89055 | 55 |
| 6 | .28864 | 3.46458 | .30764 | 3.25055 | .32685 | 3.05950 | .34628 | 2.88783 | 54 |
| 7 | .28895 | 3.46080 | .30796 | 3.24719 | .32717 | 3.05649 | .34661 | 2.88511 | 53 |
| 8 | .28927 | 3.45703 | .30828 | 3.24383 | .32749 | 3.05349 | .34693 | 2.88240 | 52 |
| 9 | .28958 | 3.45327 | .30860 | 3.24049 | .32782 | 3.05049 | .34726 | 2.87970 | 51 |
| 10 | .28990 | 3.44951 | .30891 | 3.23714 | .32814 | 3.04749 | .34758 | 2.87700 | 50 |
| 11 | .29021 | 3.44576 | .30923 | 3.23381 | .32846 | 3.04450 | .34791 | 2.87430 | 49 |
| 12 | .29053 | 3.44202 | .30955 | 3.23048 | .32878 | 3.04152 | .34824 | 2.87161 | 48 |
| 13 | .29084 | 3.43829 | .30987 | 3.22715 | .32911 | 3.03854 | .34856 | 2.86892 | 47 |
| 14 | .29116 | 3.43456 | .31019 | 3.22384 | .32943 | 3.03556 | .34889 | 2.86624 | 46 |
| 15 | .29147 | 3.43084 | .31051 | 3.22053 | .32975 | 3.03260 | .34922 | 2.86356 | 45 |
| 16 | .29179 | 3.42713 | .31083 | 3.21722 | .33007 | 3.02963 | .34954 | 2.86089 | 44 |
| 17 | .29210 | 3.42343 | .31115 | 3.21392 | .33040 | 3.02667 | .34987 | 2.85822 | 43 |
| 18 | .29242 | 3.41973 | .31147 | 3.21063 | .33072 | 3.02372 | .35020 | 2.85555 | 42 |
| 19 | .29274 | 3.41604 | .31178 | 3.20734 | .33104 | 3.02077 | .35052 | 2.85289 | 41 |
| 20 | .29305 | 3.41236 | .31210 | 3.20406 | .33136 | 3.01783 | .35085 | 2.85023 | 40 |
| 21 | .29337 | 3.40869 | .31242 | 3.20079 | .33169 | 3.01489 | .35118 | 2.84758 | 39 |
| 22 | .29368 | 3.40502 | .31274 | 3.19752 | .33201 | 3.01196 | .35150 | 2.84494 | 38 |
| 23 | .29400 | 3.40136 | .31306 | 3.19426 | .33233 | 3.00903 | .35183 | 2.84229 | 37 |
| 24 | .29432 | 3.39771 | .31338 | 3.19100 | .33266 | 3.00611 | .35216 | 2.83965 | 36 |
| 25 | .29463 | 3.39406 | .31370 | 3.18775 | .33298 | 3.00319 | .35248 | 2.83702 | 35 |
| 26 | .29495 | 3.39042 | .31402 | 3.18451 | .33330 | 3.00028 | .35281 | 2.83439 | 34 |
| 27 | .29526 | 3.38679 | .31434 | 3.18127 | .33363 | 2.99738 | .35314 | 2.83176 | 33 |
| 28 | .29558 | 3.38317 | .31466 | 3.17804 | .33395 | 2.99447 | .35346 | 2.82914 | 32 |
| 29 | .29590 | 3.37955 | .31498 | 3.17481 | .33427 | 2.99158 | .35379 | 2.82653 | 31 |
| 30 | .29621 | 3.37594 | .31530 | 3.17159 | .33460 | 2.98868 | .35412 | 2.82391 | 30 |
| 31 | .29653 | 3.37234 | .31562 | 3.16838 | .33492 | 2.98580 | .35445 | 2.82130 | 29 |
| 32 | .29685 | 3.36875 | .31594 | 3.16517 | .33524 | 2.98292 | .35477 | 2.81870 | 28 |
| 33 | .29716 | 3.36516 | .31626 | 3.16197 | .33557 | 2.98004 | .35510 | 2.81610 | 27 |
| 34 | .29748 | 3.36158 | .31658 | 3.15877 | .33589 | 2.97717 | .35543 | 2.81350 | 26 |
| 35 | .29780 | 3.35800 | .31690 | 3.15558 | .33621 | 2.97430 | .35576 | 2.81091 | 25 |
| 36 | .29811 | 3.35443 | .31722 | 3.15240 | .33654 | 2.97144 | .35608 | 2.80833 | 24 |
| 37 | .29843 | 3.35087 | .31754 | 3.14922 | .33686 | 2.96858 | .35641 | 2.80574 | 23 |
| 38 | .29875 | 3.34732 | .31786 | 3.14605 | .33718 | 2.96573 | .35674 | 2.80316 | 22 |
| 39 | .29906 | 3.34377 | .31818 | 3.14288 | .33751 | 2.96288 | .35707 | 2.80059 | 21 |
| 40 | .29938 | 3.34023 | .31850 | 3.13972 | .33783 | 2.96004 | .35740 | 2.79802 | 20 |
| 41 | .29970 | 3.33670 | .31882 | 3.13656 | .33816 | 2.95721 | .35772 | 2.79545 | 19 |
| 42 | .30001 | 3.33317 | .31914 | 3.13341 | .33848 | 2.95437 | .35805 | 2.79289 | 18 |
| 43 | .30033 | 3.32965 | .31946 | 3.13027 | .33881 | 2.95155 | .35838 | 2.79033 | 17 |
| 44 | .30065 | 3.32614 | .31978 | 3.12718 | .33913 | 2.94872 | .35871 | 2.78778 | 16 |
| 45 | .30097 | 3.32264 | .32010 | 3.12400 | .33945 | 2.94591 | .35904 | 2.78523 | 15 |
| 46 | .30128 | 3.31914 | .32042 | 3.12087 | .33978 | 2.94309 | .35937 | 2.78269 | 14 |
| 47 | .30160 | 3.31565 | .32074 | 3.11775 | .34010 | 2.94028 | .35969 | 2.78014 | 13 |
| 48 | .30192 | 3.31216 | .32106 | 3.11464 | .34043 | 2.93748 | .36002 | 2.77761 | 12 |
| 49 | .30224 | 3.30868 | .32139 | 3.11153 | .34075 | 2.93468 | .36035 | 2.77507 | 11 |
| 50 | .30255 | 3.30521 | .32171 | 3.10842 | .34108 | 2.93189 | .36068 | 2.77254 | 10 |
| 51 | .30287 | 3.30174 | .32203 | 3.10532 | .34140 | 2.92910 | .36101 | 2.77002 | 9 |
| 52 | .30319 | 3.29829 | .32235 | 3.10223 | .34173 | 2.92632 | .36134 | 2.76750 | 8 |
| 53 | .30351 | 3.29483 | .32267 | 3.09914 | .34205 | 2.92354 | .36167 | 2.76498 | 7 |
| 54 | .30382 | 3.29139 | .32299 | 3.09606 | .34238 | 2.92076 | .36199 | 2.76247 | 6 |
| 55 | .30414 | 3.28795 | .32331 | 3.09298 | .34270 | 2.91799 | .36232 | 2.75996 | 5 |
| 56 | .30446 | 3.28452 | .32363 | 3.08991 | .34303 | 2.91523 | .36265 | 2.75746 | 4 |
| 57 | .30478 | 3.28109 | .32396 | 3.08685 | .34335 | 2.91246 | .36298 | 2.75496 | 3 |
| 58 | .30509 | 3.27767 | .32428 | 3.08379 | .34368 | 2.90971 | .36331 | 2.75246 | 2 |
| 59 | .30541 | 3.27426 | .32460 | 3.08073 | .34400 | 2.90696 | .36364 | 2.74997 | 1 |
| 60 | .30573 | 3.27085 | .32492 | 3.07768 | .34433 | 2.90421 | .36397 | 2.74748 | 0 |
| | Cotang | Tang | Cotang | Tang | Cotang | Tang | Cotang | Tang | ′ |
| | 73° | | 72° | | 71° | | 70° | | |

## NATURAL TANGENTS AND COTANGENTS.

| ′ | 20° | | 21° | | 22° | | 23° | | ′ |
|---|---|---|---|---|---|---|---|---|---|
| | Tang | Cotang | Tang | Cotang | Tang | Cotang | Tang | Cotang | |
| 0 | .36397 | 2.74748 | .38386 | 2.60509 | .40403 | 2.47509 | .42447 | 2.35585 | 60 |
| 1 | .36430 | 2.74499 | .38420 | 2.60283 | .40436 | 2.47302 | .42482 | 2.35395 | 59 |
| 2 | .36463 | 2.74251 | .38453 | 2.60057 | .40470 | 2.47095 | .42516 | 2.35205 | 58 |
| 3 | .36496 | 2.74004 | .38487 | 2.59831 | .40504 | 2.46888 | .42551 | 2.35015 | 57 |
| 4 | .36529 | 2.73756 | .38520 | 2.59606 | .40538 | 2.46682 | .42585 | 2.34825 | 56 |
| 5 | .36562 | 2.73509 | .38553 | 2.59381 | .40572 | 2.46476 | .42619 | 2.34636 | 55 |
| 6 | .36595 | 2.73263 | .38587 | 2.59156 | .40606 | 2.46270 | .42654 | 2.34447 | 54 |
| 7 | .36628 | 2.73017 | .38620 | 2.58932 | .40640 | 2.46065 | .42688 | 2.34258 | 53 |
| 8 | .36661 | 2.72771 | .38654 | 2.58708 | .40674 | 2.45860 | .42722 | 2.34069 | 52 |
| 9 | .36694 | 2.72526 | .38687 | 2.58484 | .40707 | 2.45655 | .42757 | 2.33881 | 51 |
| 10 | .36727 | 2.72281 | .38721 | 2.58261 | .40741 | 2.45451 | .42791 | 2.33693 | 50 |
| 11 | .36760 | 2.72036 | .38754 | 2.58038 | .40775 | 2.45246 | .42826 | 2.33505 | 49 |
| 12 | .36793 | 2.71792 | .38787 | 2.57815 | .40809 | 2.45043 | .42860 | 2.33317 | 48 |
| 13 | .36826 | 2.71548 | .38821 | 2.57593 | .40843 | 2.44839 | .42894 | 2.33130 | 47 |
| 14 | .36859 | 2.71305 | .38854 | 2.57371 | .40877 | 2.44636 | .42929 | 2.32943 | 46 |
| 15 | .36892 | 2.71062 | .38888 | 2.57150 | .40911 | 2.44433 | .42963 | 2.32756 | 45 |
| 16 | .36925 | 2.70819 | .38921 | 2.56928 | .40945 | 2.44230 | .42998 | 2.32570 | 44 |
| 17 | .36958 | 2.70577 | .38955 | 2.56707 | .40979 | 2.44027 | .43032 | 2.32383 | 43 |
| 18 | .36991 | 2.70335 | .38988 | 2.56487 | .41013 | 2.43825 | .43067 | 2.32197 | 42 |
| 19 | .37024 | 2.70094 | .39022 | 2.56266 | .41047 | 2.43623 | .43101 | 2.32012 | 41 |
| 20 | .37057 | 2.69853 | .39055 | 2.56046 | .41081 | 2.43422 | .43136 | 2.31826 | 40 |
| 21 | .37090 | 2.69612 | .39089 | 2.55827 | .41115 | 2.43220 | .43170 | 2.31641 | 39 |
| 22 | .37123 | 2.69371 | .39122 | 2.55608 | .41149 | 2.43019 | .43205 | 2.31456 | 38 |
| 23 | .37157 | 2.69131 | .39156 | 2.55389 | .41183 | 2.42819 | .43239 | 2.31271 | 37 |
| 24 | .37190 | 2.68892 | .39190 | 2.55170 | .41217 | 2.42618 | .43274 | 2.31086 | 36 |
| 25 | .37223 | 2.68653 | .39223 | 2.54952 | .41251 | 2.42418 | .43308 | 2.30902 | 35 |
| 26 | .37256 | 2.68414 | .39257 | 2.54734 | .41285 | 2.42218 | .43343 | 2.30718 | 34 |
| 27 | .37289 | 2.68175 | .39290 | 2.54516 | .41319 | 2.42019 | .43378 | 2.30534 | 33 |
| 28 | .37322 | 2.67937 | .39324 | 2.54299 | .41353 | 2.41819 | .43412 | 2.30351 | 32 |
| 29 | .37355 | 2.67700 | .39357 | 2.54082 | .41387 | 2.41620 | .43447 | 2.30167 | 31 |
| 30 | .37388 | 2.67462 | .39391 | 2.53865 | .41421 | 2.41421 | .43481 | 2.29984 | 30 |
| 31 | .37422 | 2.67225 | .39425 | 2.53648 | .41455 | 2.41223 | .43516 | 2.29801 | 29 |
| 32 | .37455 | 2.66989 | .39458 | 2.53432 | .41490 | 2.41025 | .43550 | 2.29619 | 28 |
| 33 | .37488 | 2.66752 | .39492 | 2.53217 | .41524 | 2.40827 | .43585 | 2.29437 | 27 |
| 34 | .37521 | 2.66516 | .39526 | 2.53001 | .41558 | 2.40629 | .43620 | 2.29254 | 26 |
| 35 | .37554 | 2.66281 | .39559 | 2.52786 | .41592 | 2.40432 | .43654 | 2.29073 | 25 |
| 36 | .37588 | 2.66046 | .39593 | 2.52571 | .41626 | 2.40235 | .43689 | 2.28891 | 24 |
| 37 | .37621 | 2.65811 | .39626 | 2.52357 | .41660 | 2.40038 | .43724 | 2.28710 | 23 |
| 38 | .37654 | 2.65576 | .39660 | 2.52142 | .41694 | 2.39841 | .43758 | 2.28528 | 22 |
| 39 | .37687 | 2.65342 | .39694 | 2.51929 | .41728 | 2.39645 | .43793 | 2.28348 | 21 |
| 40 | .37720 | 2.65109 | .39727 | 2.51715 | .41763 | 2.39449 | .43828 | 2.28167 | 20 |
| 41 | .37754 | 2.64875 | .39761 | 2.51502 | .41797 | 2.39253 | .43862 | 2.27987 | 19 |
| 42 | .37787 | 2.64642 | .39795 | 2.51289 | .41831 | 2.39058 | .43897 | 2.27806 | 18 |
| 43 | .37820 | 2.64410 | .39829 | 2.51076 | .41865 | 2.38863 | .43932 | 2.27626 | 17 |
| 44 | .37853 | 2.64177 | .39862 | 2.50864 | .41899 | 2.38668 | .43966 | 2.27447 | 16 |
| 45 | .37887 | 2.63945 | .39896 | 2.50652 | .41933 | 2.38473 | .44001 | 2.27267 | 15 |
| 46 | .37920 | 2.63714 | .39930 | 2.50440 | .41968 | 2.38279 | .44036 | 2.27088 | 14 |
| 47 | .37953 | 2.63483 | .39963 | 2.50229 | .42002 | 2.38084 | .44071 | 2.26909 | 13 |
| 48 | .37986 | 2.63252 | .39997 | 2.50018 | .42036 | 2.37891 | .44105 | 2.26730 | 12 |
| 49 | .38020 | 2.63021 | .40031 | 2.49807 | .42070 | 2.37697 | .44140 | 2.26552 | 11 |
| 50 | .38053 | 2.62791 | .40065 | 2.49597 | .42105 | 2.37504 | .44175 | 2.26374 | 10 |
| 51 | .38086 | 2.62561 | .40098 | 2.49386 | .42139 | 2.37311 | .44210 | 2.26196 | 9 |
| 52 | .38120 | 2.62332 | .40132 | 2.49177 | .42173 | 2.37118 | .44244 | 2.26018 | 8 |
| 53 | .38153 | 2.62103 | .40166 | 2.48967 | .42207 | 2.36925 | .44279 | 2.25840 | 7 |
| 54 | .38186 | 2.61874 | .40200 | 2.48758 | .42242 | 2.36733 | .44314 | 2.25663 | 6 |
| 55 | .38220 | 2.61646 | .40234 | 2.48549 | .42276 | 2.36541 | .44349 | 2.25486 | 5 |
| 56 | .38253 | 2.61418 | .40267 | 2.48340 | .42310 | 2.36349 | .44384 | 2.25309 | 4 |
| 57 | .38286 | 2.61190 | .40301 | 2.48132 | .42345 | 2.36158 | .44418 | 2.25132 | 3 |
| 58 | .38320 | 2.60963 | .40335 | 2.47924 | .42379 | 2.35967 | .44453 | 2.24956 | 2 |
| 59 | .38353 | 2.60736 | .40369 | 2.47716 | .42413 | 2.35776 | .44488 | 2.24780 | 1 |
| 60 | .38386 | 2.60509 | .40403 | 2.47509 | .42447 | 2.35585 | .44523 | 2.24604 | 0 |
| ′ | Cotang | Tang | Cotang | Tang | Cotang | Tang | Cotang | Tang | ′ |
| | 69° | | 68° | | 67° | | 66° | | |

## NATURAL TANGENTS AND COTANGENTS.

| ′ | 24° Tang | 24° Cotang | 25° Tang | 25° Cotang | 26° Tang | 26° Cotang | 27° Tang | 27° Cotang | ′ |
|---|---|---|---|---|---|---|---|---|---|
| 0 | .44523 | 2.24604 | .46631 | 2.14451 | .48773 | 2.05030 | .50953 | 1.96261 | 60 |
| 1 | .44558 | 2.24528 | .46666 | 2.14288 | .48809 | 2.04879 | .50989 | 1.96120 | 59 |
| 2 | .44593 | 2.24252 | .46702 | 2.14125 | .48845 | 2.04728 | .51026 | 1.95979 | 58 |
| 3 | .44627 | 2.24077 | .46737 | 2.13963 | .48881 | 2.04577 | .51063 | 1.95838 | 57 |
| 4 | .44662 | 2.23902 | .46772 | 2.13801 | .48917 | 2.04426 | .51099 | 1.95698 | 56 |
| 5 | .44697 | 2.23727 | .46808 | 2.13639 | .48953 | 2.04276 | .51136 | 1.95557 | 55 |
| 6 | .44732 | 2.23553 | .46843 | 2.13477 | .48989 | 2.04125 | .51173 | 1.95417 | 54 |
| 7 | .44767 | 2.23378 | .46879 | 2.13316 | .49026 | 2.03975 | .51209 | 1.95277 | 53 |
| 8 | .44802 | 2.23204 | .46914 | 2.13154 | .49062 | 2.03825 | .51246 | 1.95137 | 52 |
| 9 | .44837 | 2.23030 | .46950 | 2.12993 | .49098 | 2.03675 | .51283 | 1.94997 | 51 |
| 10 | .44872 | 2.22857 | .46985 | 2.12832 | .49134 | 2.03526 | .51319 | 1.94858 | 50 |
| 11 | .44907 | 2.22683 | .47021 | 2.12671 | .49170 | 2.03376 | .51356 | 1.94718 | 49 |
| 12 | .44942 | 2.22510 | .47056 | 2.12511 | .49206 | 2.03227 | .51393 | 1.94579 | 48 |
| 13 | .44977 | 2.22337 | .47092 | 2.12350 | .49242 | 2.03078 | .51430 | 1.94440 | 47 |
| 14 | .45012 | 2.22164 | .47128 | 2.12190 | .49278 | 2.02929 | .51467 | 1.94301 | 46 |
| 15 | .45047 | 2.21992 | .47163 | 2.12030 | .49315 | 2.02780 | .51503 | 1.94162 | 45 |
| 16 | .45082 | 2.21819 | .47199 | 2.11871 | .49351 | 2.02631 | .51540 | 1.94023 | 44 |
| 17 | .45117 | 2.21647 | .47234 | 2.11711 | .49387 | 2.02483 | .51577 | 1.93885 | 43 |
| 18 | .45152 | 2.21475 | .47270 | 2.11552 | .49423 | 2.02335 | .51614 | 1.93746 | 42 |
| 19 | .45187 | 2.21304 | .47305 | 2.11392 | .49459 | 2.02187 | .51651 | 1.93608 | 41 |
| 20 | .45222 | 2.21132 | .47341 | 2.11233 | .49495 | 2.02039 | .51688 | 1.93470 | 40 |
| 21 | .45257 | 2.20961 | .47377 | 2.11075 | .49532 | 2.01891 | .51724 | 1.93332 | 39 |
| 22 | .45292 | 2.20790 | .47412 | 2.10916 | .49568 | 2.01743 | .51761 | 1.93195 | 38 |
| 23 | .45327 | 2.20619 | .47448 | 2.10758 | .49604 | 2.01596 | .51798 | 1.93057 | 37 |
| 24 | .45362 | 2.20449 | .47483 | 2.10600 | .49640 | 2.01449 | .51835 | 1.92920 | 36 |
| 25 | .45397 | 2.20278 | .47519 | 2.10442 | .49677 | 2.01302 | .51872 | 1.92782 | 35 |
| 26 | .45432 | 2.20108 | .47555 | 2.10284 | .49713 | 2.01155 | .51909 | 1.92645 | 34 |
| 27 | .45467 | 2.19938 | .47590 | 2.10126 | .49749 | 2.01008 | .51946 | 1.92508 | 33 |
| 28 | .45502 | 2.19769 | .47626 | 2.09969 | .49786 | 2.00862 | .51983 | 1.92371 | 32 |
| 29 | .45538 | 2.19599 | .47662 | 2.09811 | .49822 | 2.00715 | .52020 | 1.92235 | 31 |
| 30 | .45573 | 2.19430 | .47698 | 2.09654 | .49858 | 2.00569 | .52057 | 1.92098 | 30 |
| 31 | .45608 | 2.19261 | .47733 | 2.09498 | .49894 | 2.00423 | .52094 | 1.91962 | 29 |
| 32 | .45643 | 2.19092 | .47769 | 2.09341 | .49931 | 2.00277 | .52131 | 1.91826 | 28 |
| 33 | .45678 | 2.18923 | .47805 | 2.09184 | .49967 | 2.00131 | .52168 | 1.91690 | 27 |
| 34 | .45713 | 2.18755 | .47840 | 2.09028 | .50004 | 1.99986 | .52205 | 1.91554 | 26 |
| 35 | .45748 | 2.18587 | .47876 | 2.08872 | .50040 | 1.99841 | .52242 | 1.91418 | 25 |
| 36 | .45784 | 2.18419 | .47912 | 2.08716 | .50076 | 1.99695 | .52279 | 1.91282 | 24 |
| 37 | .45819 | 2.18251 | .47948 | 2.08560 | .50113 | 1.99550 | .52316 | 1.91147 | 23 |
| 38 | .45854 | 2.18084 | .47984 | 2.08405 | .50149 | 1.99406 | .52353 | 1.91012 | 22 |
| 39 | .45889 | 2.17916 | .48019 | 2.08250 | .50185 | 1.99261 | .52390 | 1.90876 | 21 |
| 40 | .45924 | 2.17749 | .48055 | 2.08094 | .50222 | 1.99116 | .52427 | 1.90741 | 20 |
| 41 | .45960 | 2.17582 | .48091 | 2.07939 | .50258 | 1.98972 | .52464 | 1.90607 | 19 |
| 42 | .45995 | 2.17416 | .48127 | 2.07785 | .50295 | 1.98828 | .52501 | 1.90472 | 18 |
| 43 | .46030 | 2.17249 | .48163 | 2.07630 | .50331 | 1.98684 | .52538 | 1.90337 | 17 |
| 44 | .46065 | 2.17083 | .48198 | 2.07476 | .50368 | 1.98540 | .52575 | 1.90203 | 16 |
| 45 | .46101 | 2.16917 | .48234 | 2.07321 | .50404 | 1.98396 | .52613 | 1.90069 | 15 |
| 46 | .46136 | 2.16751 | .48270 | 2.07167 | .50441 | 1.98253 | .52650 | 1.89935 | 14 |
| 47 | .46171 | 2.16585 | .48306 | 2.07014 | .50477 | 1.98110 | .52687 | 1.89801 | 13 |
| 48 | .46206 | 2.16420 | .48342 | 2.06860 | .50514 | 1.97966 | .52724 | 1.89667 | 12 |
| 49 | .46242 | 2.16255 | .48378 | 2.06706 | .50550 | 1.97823 | .52761 | 1.89533 | 11 |
| 50 | .46277 | 2.16090 | .48414 | 2.06553 | .50587 | 1.97681 | .52798 | 1.89400 | 10 |
| 51 | .46312 | 2.15925 | .48450 | 2.06400 | .50623 | 1.97538 | .52836 | 1.89266 | 9 |
| 52 | .46348 | 2.15760 | .48486 | 2.06247 | .50660 | 1.97395 | .52873 | 1.89133 | 8 |
| 53 | .46383 | 2.15596 | .48521 | 2.06094 | .50696 | 1.97253 | .52910 | 1.89000 | 7 |
| 54 | .46418 | 2.15432 | .48557 | 2.05942 | .50733 | 1.97111 | .52947 | 1.88867 | 6 |
| 55 | .46454 | 2.15268 | .48593 | 2.05790 | .50769 | 1.96969 | .52985 | 1.88734 | 5 |
| 56 | .46489 | 2.15104 | .48629 | 2.05637 | .50806 | 1.96827 | .53022 | 1.88602 | 4 |
| 57 | .46525 | 2.14940 | .48665 | 2.05485 | .50843 | 1.96685 | .53059 | 1.88469 | 3 |
| 58 | .46560 | 2.14777 | .48701 | 2.05333 | .50879 | 1.96544 | .53096 | 1.88337 | 2 |
| 59 | .46595 | 2.14614 | .48737 | 2.05182 | .50916 | 1.96402 | .53134 | 1.88205 | 1 |
| 60 | .46631 | 2.14451 | .48773 | 2.05030 | .50953 | 1.96261 | .53171 | 1.88073 | 0 |
| ′ | Cotang | Tang | Cotang | Tang | Cotang | Tang | Cotang | Tang | ′ |
|   | 65° |   | 64° |   | 63° |   | 62° |   |   |

## NATURAL TANGENTS AND COTANGENTS.

| ′ | 28° Tang | 28° Cotang | 29° Tang | 29° Cotang | 30° Tang | 30° Cotang | 31° Tang | 31° Cotang | ′ |
|---|---|---|---|---|---|---|---|---|---|
| 0  | .53171 | 1.88073 | .55431 | 1.80405 | .57735 | 1.73205 | .60086 | 1.66428 | 60 |
| 1  | .53208 | 1.87941 | .55469 | 1.80281 | .57774 | 1.73089 | .60126 | 1.66318 | 59 |
| 2  | .53246 | 1.87809 | .55507 | 1.80158 | .57813 | 1.72972 | .60165 | 1.66209 | 58 |
| 3  | .53283 | 1.87677 | .55545 | 1.80034 | .57851 | 1.72857 | .60205 | 1.66099 | 57 |
| 4  | .53320 | 1.87546 | .55583 | 1.79911 | .57890 | 1.72741 | .60245 | 1.65990 | 56 |
| 5  | .53358 | 1.87415 | .55621 | 1.79788 | .57929 | 1.72625 | .60284 | 1.65881 | 55 |
| 6  | .53395 | 1.87283 | .55659 | 1.79665 | .57968 | 1.72509 | .60324 | 1.65772 | 54 |
| 7  | .53432 | 1.87152 | .55697 | 1.79542 | .58007 | 1.72393 | .60364 | 1.65663 | 53 |
| 8  | .53470 | 1.87021 | .55736 | 1.79419 | .58046 | 1.72278 | .60403 | 1.65554 | 52 |
| 9  | .53507 | 1.86891 | .55774 | 1.79296 | .58085 | 1.72163 | .60443 | 1.65445 | 51 |
| 10 | .53545 | 1.86760 | .55812 | 1.79174 | .58124 | 1.72047 | .60483 | 1.65337 | 50 |
| 11 | .53582 | 1.86630 | .55850 | 1.79051 | .58162 | 1.71932 | .60522 | 1.65228 | 49 |
| 12 | .53620 | 1.86499 | .55888 | 1.78929 | .58201 | 1.71817 | .60562 | 1.65120 | 48 |
| 13 | .53657 | 1.86369 | .55926 | 1.78807 | .58240 | 1.71702 | .60602 | 1.65011 | 47 |
| 14 | .53694 | 1.86239 | .55964 | 1.78685 | .58279 | 1.71588 | .60642 | 1.64903 | 46 |
| 15 | .53732 | 1.86109 | .56003 | 1.78563 | .58318 | 1.71473 | .60681 | 1.64795 | 45 |
| 16 | .53769 | 1.85979 | .56041 | 1.78441 | .58357 | 1.71358 | .60721 | 1.64687 | 44 |
| 17 | .53807 | 1.85850 | .56079 | 1.78319 | .58396 | 1.71244 | .60761 | 1.64579 | 43 |
| 18 | .53844 | 1.85720 | .56117 | 1.78198 | .58435 | 1.71129 | .60801 | 1.64471 | 42 |
| 19 | .53882 | 1.85591 | .56156 | 1.78077 | .58474 | 1.71015 | .60841 | 1.64363 | 41 |
| 20 | .53920 | 1.85462 | .56194 | 1.77955 | .58513 | 1.70901 | .60881 | 1.64256 | 40 |
| 21 | .53957 | 1.85333 | .56232 | 1.77834 | .58552 | 1.70787 | .60921 | 1.64148 | 39 |
| 22 | .53995 | 1.85204 | .56270 | 1.77713 | .58591 | 1.70673 | .60960 | 1.64041 | 38 |
| 23 | .54032 | 1.85075 | .56309 | 1.77592 | .58631 | 1.70560 | .61000 | 1.63934 | 37 |
| 24 | .54070 | 1.84946 | .56347 | 1.77471 | .58670 | 1.70446 | .61040 | 1.63826 | 36 |
| 25 | .54107 | 1.84818 | .56385 | 1.77351 | .58709 | 1.70332 | .61080 | 1.63719 | 35 |
| 26 | .54145 | 1.84689 | .56424 | 1.77230 | .58748 | 1.70219 | .61120 | 1.63612 | 34 |
| 27 | .54183 | 1.84561 | .56462 | 1.77110 | .58787 | 1.70106 | .61160 | 1.63505 | 33 |
| 28 | .54220 | 1.84433 | .56501 | 1.76990 | .58826 | 1.69992 | .61200 | 1.63398 | 32 |
| 29 | .54258 | 1.84305 | .56539 | 1.76869 | .58865 | 1.69879 | .61240 | 1.63292 | 31 |
| 30 | .54296 | 1.84177 | .56577 | 1.76749 | .58905 | 1.69766 | .61280 | 1.63185 | 30 |
| 31 | .54333 | 1.84049 | .56616 | 1.76629 | .58944 | 1.69653 | .61320 | 1.63079 | 29 |
| 32 | .54371 | 1.83922 | .56654 | 1.76510 | .58983 | 1.69541 | .61360 | 1.62972 | 28 |
| 33 | .54409 | 1.83794 | .56693 | 1.76390 | .59022 | 1.69428 | .61400 | 1.62866 | 27 |
| 34 | .54446 | 1.83667 | .56731 | 1.76271 | .59061 | 1.69316 | .61440 | 1.62760 | 26 |
| 35 | .54484 | 1.83540 | .56769 | 1.76151 | .59101 | 1.69203 | .61480 | 1.62654 | 25 |
| 36 | .54522 | 1.83413 | .56808 | 1.76032 | .59140 | 1.69091 | .61520 | 1.62548 | 24 |
| 37 | .54560 | 1.83286 | .56846 | 1.75913 | .59179 | 1.68979 | .61561 | 1.62442 | 23 |
| 38 | .54597 | 1.83159 | .56885 | 1.75794 | .59218 | 1.68866 | .61601 | 1.62336 | 22 |
| 39 | .54635 | 1.83033 | .56923 | 1.75675 | .59258 | 1.68754 | .61641 | 1.62230 | 21 |
| 40 | .54673 | 1.82906 | .56962 | 1.75556 | .59297 | 1.68643 | .61681 | 1.62125 | 20 |
| 41 | .54711 | 1.82780 | .57000 | 1.75437 | .59336 | 1.68531 | .61721 | 1.62019 | 19 |
| 42 | .54748 | 1.82654 | .57039 | 1.75319 | .59376 | 1.68419 | .61761 | 1.61914 | 18 |
| 43 | .54786 | 1.82528 | .57078 | 1.75200 | .59415 | 1.68308 | .61801 | 1.61808 | 17 |
| 44 | .54824 | 1.82402 | .57116 | 1.75082 | .59454 | 1.68196 | .61842 | 1.61703 | 16 |
| 45 | .54862 | 1.82276 | .57155 | 1.74964 | .59494 | 1.68085 | .61882 | 1.61598 | 15 |
| 46 | .54900 | 1.82150 | .57193 | 1.74846 | .59533 | 1.67974 | .61922 | 1.61493 | 14 |
| 47 | .54938 | 1.82025 | .57232 | 1.74728 | .59573 | 1.67863 | .61962 | 1.61388 | 13 |
| 48 | .54975 | 1.81899 | .57271 | 1.74610 | .59612 | 1.67752 | .62003 | 1.61283 | 12 |
| 49 | .55013 | 1.81774 | .57309 | 1.74492 | .59651 | 1.67641 | .62043 | 1.61179 | 11 |
| 50 | .55051 | 1.81649 | .57348 | 1.74375 | .59691 | 1.67530 | .62083 | 1.61074 | 10 |
| 51 | .55089 | 1.81524 | .57386 | 1.74257 | .59730 | 1.67419 | .62124 | 1.60970 | 9 |
| 52 | .55127 | 1.81399 | .57425 | 1.74140 | .59770 | 1.67309 | .62164 | 1.60865 | 8 |
| 53 | .55165 | 1.81274 | .57464 | 1.74022 | .59809 | 1.67198 | .62204 | 1.60761 | 7 |
| 54 | .55203 | 1.81150 | .57503 | 1.73905 | .59849 | 1.67088 | .62245 | 1.60657 | 6 |
| 55 | .55241 | 1.81025 | .57541 | 1.73788 | .59888 | 1.66978 | .62285 | 1.60553 | 5 |
| 56 | .55279 | 1.80901 | .57580 | 1.73671 | .59928 | 1.66867 | .62325 | 1.60449 | 4 |
| 57 | .55317 | 1.80777 | .57619 | 1.73555 | .59967 | 1.66757 | .62366 | 1.60345 | 3 |
| 58 | .55355 | 1.80653 | .57657 | 1.73438 | .60007 | 1.66647 | .62406 | 1.60241 | 2 |
| 59 | .55393 | 1.80529 | .57696 | 1.73321 | .60046 | 1.66538 | .62446 | 1.60137 | 1 |
| 60 | .55431 | 1.80405 | .57735 | 1.73205 | .60086 | 1.66428 | .62487 | 1.60033 | 0 |
| ′ | Cotang | Tang | Cotang | Tang | Cotang | Tang | Cotang | Tang | ′ |
|   | 61°    |      | 60°    |      | 59°    |      | 58°    |      |   |

## NATURAL TANGENTS AND COTANGENTS.    129

| ′ | 32° Tang | 32° Cotang | 33° Tang | 33° Cotang | 34° Tang | 34° Cotang | 35° Tang | 35° Cotang | ′ |
|---|---|---|---|---|---|---|---|---|---|
| 0 | .62487 | 1.60033 | .64941 | 1.53986 | .67451 | 1.48256 | .70021 | 1.42815 | 60 |
| 1 | .62527 | 1.59930 | .64982 | 1.53888 | .67493 | 1.48163 | .70064 | 1.42726 | 59 |
| 2 | .62568 | 1.59826 | .65024 | 1.53791 | .67536 | 1.48070 | .70107 | 1.42638 | 58 |
| 3 | .62608 | 1.59723 | .65065 | 1.53693 | .67578 | 1.47977 | .70151 | 1.42550 | 57 |
| 4 | .62649 | 1.59620 | .65106 | 1.53595 | .67620 | 1.47885 | .70194 | 1.42462 | 56 |
| 5 | .62689 | 1.59517 | .65148 | 1.53497 | .67663 | 1.47792 | .70238 | 1.42374 | 55 |
| 6 | .62730 | 1.59414 | .65189 | 1.53400 | .67705 | 1.47699 | .70281 | 1.42286 | 54 |
| 7 | .62770 | 1.59311 | .65231 | 1.53302 | .67748 | 1.47607 | .70325 | 1.42198 | 53 |
| 8 | .62811 | 1.59208 | .65272 | 1.53205 | .67790 | 1.47514 | .70368 | 1.42110 | 52 |
| 9 | .62852 | 1.59105 | .65314 | 1.53107 | .67832 | 1.47422 | .70412 | 1.42022 | 51 |
| 10 | .62892 | 1.59002 | .65355 | 1.53010 | .67875 | 1.47330 | .70455 | 1.41934 | 50 |
| 11 | .62933 | 1.58900 | .65397 | 1.52913 | .67917 | 1.47238 | .70499 | 1.41847 | 49 |
| 12 | .62973 | 1.58797 | .65438 | 1.52816 | .67960 | 1.47146 | .70542 | 1.41759 | 48 |
| 13 | .63014 | 1.58695 | .65480 | 1.52719 | .68002 | 1.47053 | .70586 | 1.41672 | 47 |
| 14 | .63055 | 1.58593 | .65521 | 1.52622 | .68045 | 1.46962 | .70629 | 1.41584 | 46 |
| 15 | .63095 | 1.58490 | .65563 | 1.52525 | .68088 | 1.46870 | .70673 | 1.41497 | 45 |
| 16 | .63136 | 1.58388 | .65604 | 1.52429 | .68130 | 1.46778 | .70717 | 1.41409 | 44 |
| 17 | .63177 | 1.58286 | .65646 | 1.52332 | .68173 | 1.46686 | .70760 | 1.41322 | 43 |
| 18 | .63217 | 1.58184 | .65688 | 1.52235 | .68215 | 1.46595 | .70804 | 1.41235 | 42 |
| 19 | .63258 | 1.58083 | .65729 | 1.52139 | .68258 | 1.46503 | .70848 | 1.41148 | 41 |
| 20 | .63299 | 1.57981 | .65771 | 1.52043 | .68301 | 1.46411 | .70891 | 1.41061 | 40 |
| 21 | .63340 | 1.57879 | .65813 | 1.51946 | .68343 | 1.46320 | .70935 | 1.40974 | 39 |
| 22 | .63380 | 1.57778 | .65854 | 1.51850 | .68386 | 1.46229 | .70979 | 1.40887 | 38 |
| 23 | .63421 | 1.57676 | .65896 | 1.51754 | .68429 | 1.46137 | .71023 | 1.40800 | 37 |
| 24 | .63462 | 1.57575 | .65938 | 1.51658 | .68471 | 1.46046 | .71066 | 1.40714 | 36 |
| 25 | .63503 | 1.57474 | .65980 | 1.51562 | .68514 | 1.45955 | .71110 | 1.40627 | 35 |
| 26 | .63544 | 1.57372 | .66021 | 1.51466 | .68557 | 1.45864 | .71154 | 1.40540 | 34 |
| 27 | .63584 | 1.57271 | .66063 | 1.51370 | .68600 | 1.45773 | .71198 | 1.40454 | 33 |
| 28 | .63625 | 1.57170 | .66105 | 1.51275 | .68642 | 1.45682 | .71242 | 1.40367 | 32 |
| 29 | .63666 | 1.57069 | .66147 | 1.51179 | .68685 | 1.45592 | .71285 | 1.40281 | 31 |
| 30 | .63707 | 1.56969 | .66189 | 1.51084 | .68728 | 1.45501 | .71329 | 1.40195 | 30 |
| 31 | .63748 | 1.56868 | .66230 | 1.50988 | .68771 | 1.45410 | .71373 | 1.40109 | 29 |
| 32 | .63789 | 1.56767 | .66272 | 1.50893 | .68814 | 1.45320 | .71417 | 1.40022 | 28 |
| 33 | .63830 | 1.56667 | .66314 | 1.50797 | .68857 | 1.45229 | .71461 | 1.39936 | 27 |
| 34 | .63871 | 1.56566 | .66356 | 1.50702 | .68900 | 1.45139 | .71505 | 1.39850 | 26 |
| 35 | .63912 | 1.56466 | .66398 | 1.50607 | .68942 | 1.45049 | .71549 | 1.39764 | 25 |
| 36 | .63953 | 1.56366 | .66440 | 1.50512 | .68985 | 1.44958 | .71593 | 1.39679 | 24 |
| 37 | .63994 | 1.56265 | .66482 | 1.50417 | .69028 | 1.44868 | .71637 | 1.39593 | 23 |
| 38 | .64035 | 1.56165 | .66524 | 1.50322 | .69071 | 1.44778 | .71681 | 1.39507 | 22 |
| 39 | .64076 | 1.56065 | .66566 | 1.50228 | .69114 | 1.44688 | .71725 | 1.39421 | 21 |
| 40 | .64117 | 1.55966 | .66608 | 1.50133 | .69157 | 1.44598 | .71769 | 1.39336 | 20 |
| 41 | .64158 | 1.55866 | .66650 | 1.50038 | .69200 | 1.44508 | .71813 | 1.39250 | 19 |
| 42 | .64199 | 1.55766 | .66692 | 1.49944 | .69243 | 1.44418 | .71857 | 1.39165 | 18 |
| 43 | .64240 | 1.55666 | .66734 | 1.49849 | .69286 | 1.44329 | .71901 | 1.39079 | 17 |
| 44 | .64281 | 1.55567 | .66776 | 1.49755 | .69329 | 1.44239 | .71946 | 1.38994 | 16 |
| 45 | .64322 | 1.55467 | .66818 | 1.49661 | .69372 | 1.44149 | .71990 | 1.38909 | 15 |
| 46 | .64363 | 1.55368 | .66860 | 1.49566 | .69416 | 1.44060 | .72034 | 1.38824 | 14 |
| 47 | .64404 | 1.55269 | .66902 | 1.49472 | .69459 | 1.43970 | .72078 | 1.38738 | 13 |
| 48 | .64446 | 1.55170 | .66944 | 1.49378 | .69502 | 1.43881 | .72122 | 1.38653 | 12 |
| 49 | .64487 | 1.55071 | .66986 | 1.49284 | .69545 | 1.43792 | .72167 | 1.38568 | 11 |
| 50 | .64528 | 1.54972 | .67028 | 1.49190 | .69588 | 1.43703 | .72211 | 1.38484 | 10 |
| 51 | .64569 | 1.54873 | .67071 | 1.49097 | .69631 | 1.43614 | .72255 | 1.38399 | 9 |
| 52 | .64610 | 1.54774 | .67113 | 1.49003 | .69675 | 1.43525 | .72299 | 1.38314 | 8 |
| 53 | .64652 | 1.54675 | .67155 | 1.48909 | .69718 | 1.43436 | .72344 | 1.38229 | 7 |
| 54 | .64693 | 1.54576 | .67197 | 1.48816 | .69761 | 1.43347 | .72388 | 1.38145 | 6 |
| 55 | .64734 | 1.54478 | .67239 | 1.48722 | .69804 | 1.43258 | .72432 | 1.38060 | 5 |
| 56 | .64775 | 1.54379 | .67282 | 1.48629 | .69847 | 1.43169 | .72477 | 1.37976 | 4 |
| 57 | .64817 | 1.54281 | .67324 | 1.48536 | .69891 | 1.43080 | .72521 | 1.37891 | 3 |
| 58 | .64858 | 1.54183 | .67366 | 1.48442 | .69934 | 1.42992 | .72565 | 1.37807 | 2 |
| 59 | .64899 | 1.54085 | .67409 | 1.48349 | .69977 | 1.42903 | .72610 | 1.37722 | 1 |
| 60 | .64941 | 1.53986 | .67451 | 1.48256 | .70021 | 1.42815 | .72654 | 1.37638 | 0 |
| ′ | Cotang | Tang | Cotang | Tang | Cotang | Tang | Cotang | Tang | ′ |
|   | 57° | | 56° | | 55° | | 54° | | |

# NATURAL TANGENTS AND COTANGENTS.

| ′ | 36° | | 37° | | 38° | | 39° | | ′ |
|---|---|---|---|---|---|---|---|---|---|
| | Tang | Cotang | Tang | Cotang | Tang | Cotang | Tang | Cotang | |
| 0 | .72654 | 1.37638 | .75355 | 1.32704 | .78129 | 1.27994 | .80978 | 1.23490 | 60 |
| 1 | .72699 | 1.37554 | .75401 | 1.32624 | .78175 | 1.27917 | .81027 | 1.23416 | 59 |
| 2 | .72743 | 1.37470 | .75447 | 1.32544 | .78222 | 1.27841 | .81075 | 1.23343 | 58 |
| 3 | .72788 | 1.37386 | .75492 | 1.32464 | .78269 | 1.27764 | .81123 | 1.23270 | 57 |
| 4 | .72832 | 1.37302 | .75538 | 1.32384 | .78316 | 1.27688 | .81171 | 1.23196 | 56 |
| 5 | .72877 | 1.37218 | .75584 | 1.32304 | .78363 | 1.27611 | .81220 | 1.23123 | 55 |
| 6 | .72921 | 1.37134 | .75629 | 1.32224 | .78410 | 1.27535 | .81268 | 1.23050 | 54 |
| 7 | .72966 | 1.37050 | .75675 | 1.32144 | .78457 | 1.27458 | .81316 | 1.22977 | 53 |
| 8 | .73010 | 1.36967 | .75721 | 1.32064 | .78504 | 1.27382 | .81364 | 1.22904 | 52 |
| 9 | .73055 | 1.36883 | .75767 | 1.31984 | .78551 | 1.27306 | .81413 | 1.22831 | 51 |
| 10 | .73100 | 1.36800 | .75812 | 1.31904 | .78598 | 1.27230 | .81461 | 1.22758 | 50 |
| 11 | .73144 | 1.36716 | .75858 | 1.31825 | .78645 | 1.27153 | .81510 | 1.22685 | 49 |
| 12 | .73189 | 1.36633 | .75904 | 1.31745 | .78692 | 1.27077 | .81558 | 1.22612 | 48 |
| 13 | .73234 | 1.36549 | .75950 | 1.31666 | .78739 | 1.27001 | .81606 | 1.22539 | 47 |
| 14 | .73278 | 1.36466 | .75996 | 1.31586 | .78786 | 1.26925 | .81655 | 1.22467 | 46 |
| 15 | .73323 | 1.36383 | .76042 | 1.31507 | .78834 | 1.26849 | .81703 | 1.22394 | 45 |
| 16 | .73368 | 1.36300 | .76088 | 1.31427 | .78881 | 1.26774 | .81752 | 1.22321 | 44 |
| 17 | .73413 | 1.36217 | .76134 | 1.31348 | .78928 | 1.26698 | .81800 | 1.22249 | 43 |
| 18 | .73457 | 1.36134 | .76180 | 1.31269 | .78975 | 1.26622 | .81849 | 1.22176 | 42 |
| 19 | .73502 | 1.36051 | .76226 | 1.31190 | .79022 | 1.26546 | .81898 | 1.22104 | 41 |
| 20 | .73547 | 1.35968 | .76272 | 1.31110 | .79070 | 1.26471 | .81946 | 1.22031 | 40 |
| 21 | .73592 | 1.35885 | .76318 | 1.31031 | .79117 | 1.26395 | .81995 | 1.21959 | 39 |
| 22 | .73637 | 1.35802 | .76364 | 1.30952 | .79164 | 1.26319 | .82044 | 1.21886 | 38 |
| 23 | .73681 | 1.35719 | .76410 | 1.30873 | .79212 | 1.26244 | .82092 | 1.21814 | 37 |
| 24 | .73726 | 1.35637 | .76456 | 1.30795 | .79259 | 1.26169 | .82141 | 1.21742 | 36 |
| 25 | .73771 | 1.35554 | .76502 | 1.30716 | .79306 | 1.26093 | .82190 | 1.21670 | 35 |
| 26 | .73816 | 1.35472 | .76548 | 1.30637 | .79354 | 1.26018 | .82238 | 1.21598 | 34 |
| 27 | .73861 | 1.35389 | .76594 | 1.30558 | .79401 | 1.25943 | .82287 | 1.21526 | 33 |
| 28 | .73906 | 1.35307 | .76640 | 1.30480 | .79449 | 1.25867 | .82336 | 1.21454 | 32 |
| 29 | .73951 | 1.35224 | .76686 | 1.30401 | .79496 | 1.25792 | .82385 | 1.21382 | 31 |
| 30 | .73996 | 1.35142 | .76733 | 1.30323 | .79544 | 1.25717 | .82434 | 1.21310 | 30 |
| 31 | .74041 | 1.35060 | .76779 | 1.30244 | .79591 | 1.25642 | .82483 | 1.21238 | 29 |
| 32 | .74086 | 1.34978 | .76825 | 1.30166 | .79639 | 1.25567 | .82531 | 1.21166 | 28 |
| 33 | .74131 | 1.34896 | .76871 | 1.30087 | .79686 | 1.25492 | .82580 | 1.21094 | 27 |
| 34 | .74176 | 1.34814 | .76918 | 1.30009 | .79734 | 1.25417 | .82629 | 1.21023 | 26 |
| 35 | .74221 | 1.34732 | .76964 | 1.29931 | .79781 | 1.25343 | .82678 | 1.20951 | 25 |
| 36 | .74267 | 1.34650 | .77010 | 1.29853 | .79829 | 1.25268 | .82727 | 1.20879 | 24 |
| 37 | .74312 | 1.34568 | .77057 | 1.29775 | .79877 | 1.25193 | .82776 | 1.20808 | 23 |
| 38 | .74357 | 1.34487 | .77103 | 1.29696 | .79924 | 1.25118 | .82825 | 1.20736 | 22 |
| 39 | .74402 | 1.34405 | .77149 | 1.29618 | .79972 | 1.25044 | .82874 | 1.20665 | 21 |
| 40 | .74447 | 1.34323 | .77196 | 1.29541 | .80020 | 1.24969 | .82923 | 1.20593 | 20 |
| 41 | .74492 | 1.34242 | .77242 | 1.29463 | .80067 | 1.24895 | .82972 | 1.20522 | 19 |
| 42 | .74538 | 1.34160 | .77289 | 1.29385 | .80115 | 1.24820 | .83022 | 1.20451 | 18 |
| 43 | .74583 | 1.34079 | .77335 | 1.29307 | .80163 | 1.24746 | .83071 | 1.20379 | 17 |
| 44 | .74628 | 1.33998 | .77382 | 1.29229 | .80211 | 1.24672 | .83120 | 1.20308 | 16 |
| 45 | .74674 | 1.33916 | .77428 | 1.29152 | .80258 | 1.24597 | .83169 | 1.20237 | 15 |
| 46 | .74719 | 1.33835 | .77475 | 1.29074 | .80306 | 1.24523 | .83218 | 1.20166 | 14 |
| 47 | .74764 | 1.33754 | .77521 | 1.28997 | .80354 | 1.24449 | .83268 | 1.20095 | 13 |
| 48 | .74810 | 1.33673 | .77568 | 1.28919 | .80402 | 1.24375 | .83317 | 1.20024 | 12 |
| 49 | .74855 | 1.33592 | .77615 | 1.28842 | .80450 | 1.24301 | .83366 | 1.19953 | 11 |
| 50 | .74900 | 1.33511 | .77661 | 1.28764 | .80498 | 1.24227 | .83415 | 1.19882 | 10 |
| 51 | .74946 | 1.33430 | .77708 | 1.28687 | .80546 | 1.24153 | .83465 | 1.19811 | 9 |
| 52 | .74991 | 1.33349 | .77754 | 1.28610 | .80594 | 1.24079 | .83514 | 1.19740 | 8 |
| 53 | .75037 | 1.33268 | .77801 | 1.28533 | .80642 | 1.24005 | .83564 | 1.19669 | 7 |
| 54 | .75082 | 1.33187 | .77848 | 1.28456 | .80690 | 1.23931 | .83613 | 1.19599 | 6 |
| 55 | .75128 | 1.33107 | .77895 | 1.28379 | .80738 | 1.23858 | .83662 | 1.19528 | 5 |
| 56 | .75173 | 1.33026 | .77941 | 1.28302 | .80786 | 1.23784 | .83712 | 1.19457 | 4 |
| 57 | .75219 | 1.32946 | .77988 | 1.28225 | .80834 | 1.23710 | .83761 | 1.19387 | 3 |
| 58 | .75264 | 1.32865 | .78035 | 1.28148 | .80882 | 1.23637 | .83811 | 1.19316 | 2 |
| 59 | .75310 | 1.32785 | .78082 | 1.28071 | .80930 | 1.23563 | .83860 | 1.19246 | 1 |
| 60 | .75355 | 1.32704 | .78129 | 1.27994 | .80978 | 1.23490 | .83910 | 1.19175 | 0 |
| ′ | Cotang | Tang | Cotang | Tang | Cotang | Tang | Cotang | Tang | ′ |
| | 53° | | 52° | | 51° | | 50° | | |

## NATURAL TANGENTS AND COTANGENTS.

| ′ | 40° | | 41° | | 42° | | 43° | | ′ |
|---|---|---|---|---|---|---|---|---|---|
| | Tang | Cotang | Tang | Cotang | Tang | Cotang | Tang | Cotang | |
| 0 | .83910 | 1.19175 | .86929 | 1.15037 | .90040 | 1.11061 | .93252 | 1.07237 | 60 |
| 1 | .83960 | 1.19105 | .86980 | 1.14969 | .90093 | 1.10996 | .93306 | 1.07174 | 59 |
| 2 | .84009 | 1.19035 | .87031 | 1.14902 | .90146 | 1.10931 | .93360 | 1.07112 | 58 |
| 3 | .84059 | 1.18964 | .87082 | 1.14834 | .90199 | 1.10867 | .93415 | 1.07049 | 57 |
| 4 | .84108 | 1.18894 | .87133 | 1.14767 | .90251 | 1.10802 | .93469 | 1.06987 | 56 |
| 5 | .84158 | 1.18824 | .87184 | 1.14699 | .90304 | 1.10737 | .93524 | 1.06925 | 55 |
| 6 | .84208 | 1.18754 | .87236 | 1.14632 | .90357 | 1.10672 | .93578 | 1.06862 | 54 |
| 7 | .84258 | 1.18684 | .87287 | 1.14565 | .90410 | 1.10607 | .93633 | 1.06800 | 53 |
| 8 | .84307 | 1.18614 | .87338 | 1.14498 | .90463 | 1.10543 | .93688 | 1.06738 | 52 |
| 9 | .84357 | 1.18544 | .87389 | 1.14430 | .90516 | 1.10478 | .93742 | 1.06676 | 51 |
| 10 | .84407 | 1.18474 | .87441 | 1.14363 | .90569 | 1.10414 | .93797 | 1.06613 | 50 |
| 11 | .84457 | 1.18404 | .87492 | 1.14296 | .90621 | 1.10349 | .93852 | 1.06551 | 49 |
| 12 | .84507 | 1.18334 | .87543 | 1.14229 | .90674 | 1.10285 | .93906 | 1.06489 | 48 |
| 13 | .84556 | 1.18264 | .87595 | 1.14162 | .90727 | 1.10220 | .93961 | 1.06427 | 47 |
| 14 | .84606 | 1.18194 | .87646 | 1.14095 | .90781 | 1.10156 | .94016 | 1.06365 | 46 |
| 15 | .84656 | 1.18125 | .87698 | 1.14028 | .90834 | 1.10091 | .94071 | 1.06303 | 45 |
| 16 | .84706 | 1.18055 | .87749 | 1.13961 | .90887 | 1.10027 | .94125 | 1.06241 | 44 |
| 17 | .84756 | 1.17986 | .87801 | 1.13894 | .90940 | 1.09963 | .94180 | 1.06179 | 43 |
| 18 | .84806 | 1.17916 | .87852 | 1.13828 | .90993 | 1.09899 | .94235 | 1.06117 | 42 |
| 19 | .84856 | 1.17846 | .87904 | 1.13761 | .91046 | 1.09834 | .94290 | 1.06056 | 41 |
| 20 | .84906 | 1.17777 | .87955 | 1.13694 | .91099 | 1.09770 | .94345 | 1.05994 | 40 |
| 21 | .84956 | 1.17708 | .88007 | 1.13627 | .91153 | 1.09706 | .94400 | 1.05932 | 39 |
| 22 | .85006 | 1.17638 | .88059 | 1.13561 | .91206 | 1.09642 | .94455 | 1.05870 | 38 |
| 23 | .85057 | 1.17569 | .88110 | 1.13494 | .91259 | 1.09578 | .94510 | 1.05809 | 37 |
| 24 | .85107 | 1.17500 | .88162 | 1.13428 | .91313 | 1.09514 | .94565 | 1.05747 | 36 |
| 25 | .85157 | 1.17430 | .88214 | 1.13361 | .91366 | 1.09450 | .94620 | 1.05685 | 35 |
| 26 | .85207 | 1.17361 | .88265 | 1.13295 | .91419 | 1.09386 | .94676 | 1.05624 | 34 |
| 27 | .85257 | 1.17292 | .88317 | 1.13228 | .91473 | 1.09322 | .94731 | 1.05562 | 33 |
| 28 | .85308 | 1.17223 | .88369 | 1.13162 | .91526 | 1.09258 | .94786 | 1.05501 | 32 |
| 29 | .85358 | 1.17154 | .88421 | 1.13096 | .91580 | 1.09195 | .94841 | 1.05439 | 31 |
| 30 | .85408 | 1.17085 | .88473 | 1.13029 | .91633 | 1.09131 | .94896 | 1.05378 | 30 |
| 31 | .85458 | 1.17016 | .88524 | 1.12963 | .91687 | 1.09067 | .94952 | 1.05317 | 29 |
| 32 | .85509 | 1.16947 | .88576 | 1.12897 | .91740 | 1.09003 | .95007 | 1.05255 | 28 |
| 33 | .85559 | 1.16878 | .88628 | 1.12831 | .91794 | 1.08940 | .95062 | 1.05194 | 27 |
| 34 | .85609 | 1.16809 | .88680 | 1.12765 | .91847 | 1.08876 | .95118 | 1.05133 | 26 |
| 35 | .85660 | 1.16741 | .88732 | 1.12699 | .91901 | 1.08813 | .95173 | 1.05072 | 25 |
| 36 | .85710 | 1.16672 | .88784 | 1.12633 | .91955 | 1.08749 | .95229 | 1.05010 | 24 |
| 37 | .85761 | 1.16603 | .88836 | 1.12567 | .92008 | 1.08686 | .95284 | 1.04949 | 23 |
| 38 | .85811 | 1.16535 | .88888 | 1.12501 | .92062 | 1.08622 | .95340 | 1.04888 | 22 |
| 39 | .85862 | 1.16466 | .88940 | 1.12435 | .92116 | 1.08559 | .95395 | 1.04827 | 21 |
| 40 | .85912 | 1.16398 | .88992 | 1.12369 | .92170 | 1.08496 | .95451 | 1.04766 | 20 |
| 41 | .85963 | 1.16329 | .89045 | 1.12303 | .92224 | 1.08432 | .95506 | 1.04705 | 19 |
| 42 | .86014 | 1.16261 | .89097 | 1.12238 | .92277 | 1.08369 | .95562 | 1.04644 | 18 |
| 43 | .86064 | 1.16192 | .89149 | 1.12172 | .92331 | 1.08306 | .95618 | 1.04583 | 17 |
| 44 | .86115 | 1.16124 | .89201 | 1.12106 | .92385 | 1.08243 | .95673 | 1.04522 | 16 |
| 45 | .86166 | 1.16056 | .89253 | 1.12041 | .92439 | 1.08179 | .95729 | 1.04461 | 15 |
| 46 | .86216 | 1.15987 | .89306 | 1.11975 | .92493 | 1.08116 | .95785 | 1.04401 | 14 |
| 47 | .86267 | 1.15919 | .89358 | 1.11909 | .92547 | 1.08053 | .95841 | 1.04340 | 13 |
| 48 | .86318 | 1.15851 | .89410 | 1.11844 | .92601 | 1.07990 | .95897 | 1.04279 | 12 |
| 49 | .86368 | 1.15783 | .89463 | 1.11778 | .92655 | 1.07927 | .95952 | 1.04218 | 11 |
| 50 | .86419 | 1.15715 | .89515 | 1.11713 | .92709 | 1.07864 | .96008 | 1.04158 | 10 |
| 51 | .86470 | 1.15647 | .89567 | 1.11648 | .92763 | 1.07801 | .96064 | 1.04097 | 9 |
| 52 | .86521 | 1.15579 | .89620 | 1.11582 | .92817 | 1.07738 | .96120 | 1.04036 | 8 |
| 53 | .86572 | 1.15511 | .89672 | 1.11517 | .92872 | 1.07676 | .96176 | 1.03976 | 7 |
| 54 | .86623 | 1.15443 | .89725 | 1.11452 | .92926 | 1.07613 | .96232 | 1.03915 | 6 |
| 55 | .86674 | 1.15375 | .89777 | 1.11387 | .92980 | 1.07550 | .96288 | 1.03855 | 5 |
| 56 | .86725 | 1.15308 | .89830 | 1.11321 | .93034 | 1.07487 | .96344 | 1.03794 | 4 |
| 57 | .86776 | 1.15240 | .89883 | 1.11256 | .93088 | 1.07425 | .96400 | 1.03734 | 3 |
| 58 | .86827 | 1.15172 | .89935 | 1.11191 | .93143 | 1.07362 | .96457 | 1.03674 | 2 |
| 59 | .86878 | 1.15104 | .89988 | 1.11126 | .93197 | 1.07299 | .96513 | 1.03613 | 1 |
| 60 | .86929 | 1.15037 | .90040 | 1.11061 | .93252 | 1.07237 | .96569 | 1.03553 | 0 |
| ′ | Cotang | Tang | Cotang | Tang | Cotang | Tang | Cotang | Tang | ′ |
| | 49° | | 48° | | 47° | | 46° | | |

# NATURAL TANGENTS AND COTANGENTS.

| ′ | 44° | | ′ | ′ | 44° | | ′ | ′ | 44° | | ′ |
|---|---|---|---|---|---|---|---|---|---|---|---|
| | Tang | Cotang | | | Tang | Cotang | | | Tang | Cotang | |
| 0 | .96569 | 1.03553 | 60 | 20 | .97700 | 1.02355 | 40 | 40 | .98843 | 1.01170 | 20 |
| 1 | .96625 | 1.03493 | 59 | 21 | .97756 | 1.02295 | 39 | 41 | .98901 | 1.01112 | 19 |
| 2 | .96681 | 1.03433 | 58 | 22 | .97813 | 1.02236 | 38 | 42 | .98958 | 1.01053 | 18 |
| 3 | .96738 | 1.03372 | 57 | 23 | .97870 | 1.02176 | 37 | 43 | .99016 | 1.00994 | 17 |
| 4 | .96794 | 1.03312 | 56 | 24 | .97927 | 1.02117 | 36 | 44 | .99073 | 1.00935 | 16 |
| 5 | .96850 | 1.03252 | 55 | 25 | .97984 | 1.02057 | 35 | 45 | .99131 | 1.00876 | 15 |
| 6 | .96907 | 1.03192 | 54 | 26 | .98041 | 1.01998 | 34 | 46 | .99189 | 1.00818 | 14 |
| 7 | .96963 | 1.03132 | 53 | 27 | .98098 | 1.01939 | 33 | 47 | .99247 | 1.00759 | 13 |
| 8 | .97020 | 1.03072 | 52 | 28 | .98155 | 1.01879 | 32 | 48 | .99304 | 1.00701 | 12 |
| 9 | .97076 | 1.03012 | 51 | 29 | .98213 | 1.01820 | 31 | 49 | .99362 | 1.00642 | 11 |
| 10 | .97133 | 1.02952 | 50 | 30 | .98270 | 1.01761 | 30 | 50 | .99420 | 1.00583 | 10 |
| 11 | .97189 | 1.02892 | 49 | 31 | .98327 | 1.01702 | 29 | 51 | .99478 | 1.00525 | 9 |
| 12 | .97246 | 1.02832 | 48 | 32 | .98384 | 1.01642 | 28 | 52 | .99536 | 1.00467 | 8 |
| 13 | .97302 | 1.02772 | 47 | 33 | .98441 | 1.01583 | 27 | 53 | .99594 | 1.00408 | 7 |
| 14 | .97359 | 1.02713 | 46 | 34 | .98499 | 1.01524 | 26 | 54 | .99652 | 1.00350 | 6 |
| 15 | .97416 | 1.02653 | 45 | 35 | .98556 | 1.01465 | 25 | 55 | .99710 | 1.00291 | 5 |
| 16 | .97472 | 1.02593 | 44 | 36 | .98613 | 1.01406 | 24 | 56 | .99768 | 1.00233 | 4 |
| 17 | .97529 | 1.02533 | 43 | 37 | .98671 | 1.01347 | 23 | 57 | .99826 | 1.00175 | 3 |
| 18 | .97586 | 1.02474 | 42 | 38 | .98728 | 1.01288 | 22 | 58 | .99884 | 1.00116 | 2 |
| 19 | .97643 | 1.02414 | 41 | 39 | .98786 | 1.01229 | 21 | 59 | .99942 | 1.00058 | 1 |
| 20 | .97700 | 1.02355 | 40 | 40 | .98843 | 1.01170 | 20 | 60 | 1.00000 | 1.00000 | 0 |
| | Cotang | Tang | | | Cotang | Tang | | | Cotang | Tang | |
| ′ | 45° | | ′ | ′ | 45° | | ′ | ′ | 45° | | ′ |

# SHORT-TITLE CATALOGUE

OF THE

## PUBLICATIONS

OF

## JOHN WILEY & SONS,

NEW YORK.

LONDON: CHAPMAN & HALL, LIMITED.

ARRANGED UNDER SUBJECTS.

---

Descriptive circulars sent on application,
Books marked with an asterisk are sold at *net* prices only.
All books are bound in cloth unless otherwise stated.

---

### AGRICULTURE.

CATTLE FEEDING—DISEASES OF ANIMALS—GARDENING, ETC.

| | | |
|---|---|---|
| Armsby's Manual of Cattle Feeding....................12mo, | $1 | 75 |
| Downing's Fruit and Fruit Trees........................8vo, | 5 | 00 |
| Kemp's Landscape Gardening.... .....................12mo, | 2 | 50 |
| Stockbridge's Rocks and Soils.... . ................ ....8vo, | 2 | 50 |
| Lloyd's Science of Agriculture..........................8vo, | 4 | 00 |
| Loudon's Gardening for Ladies. (Downing.)...........12mo, | 1 | 50 |
| Steel's Treatise on the Diseases of the Ox.................8vo, | 6 | 00 |
| " Treatise on the Diseases of the Dog................8vo, | 3 | 50 |
| Grotenfelt's The Principles of Modern Dairy Practice. (Woll.) 12mo, | 2 | 00 |

### ARCHITECTURE.

BUILDING—CARPENTRY—STAIRS, ETC.

| | | |
|---|---|---|
| Berg's Buildings and Structures of American Railroads.....4to, | 7 | 50 |
| Birkmire's Architectural Iron and Steel...................8vo, | 3 | 50 |
| " Skeleton Construction in Buildings............8vo, | 3 | 00 |

1

| | | |
|---|---:|---:|
| Birkmire's **Compound Riveted Girders** ...........8vo, | $2 | 00 |
| "       American Theatres—Planning and Construction.8vo, | 3 | 00 |
| Carpenter's Heating and Ventilating of Buildings..........8vo, | 3 | 00 |
| Freitag's **Architectural** Engineering................8vo, | 2 | 50 |
| Kidder's **Architect and** Builder's Pocket-book.....Morocco flap, | 4 | 00 |
| Hatfield's **American** House Carpenter................ 8vo, | 5 | 00 |
| "       **Transverse Strains**...........................8vo, | 5 | 00 |
| Monckton's **Stair Building—Wood, Iron, and Stone**.......4to, | 4 | 00 |
| Gerhard's Sanitary House Inspection...................16mo, | 1 | 00 |
| Downing and Wightwick's **Hints to** Architects..........8vo, | 2 | 00 |
| "       Cottages........................................8vo, | 2 | 50 |
| Holly's Carpenter **and Joiner**.......................18mo, | | 75 |
| Worcester's Small Hospitals—**Establishment and Maintenance,** including Atkinson's **Suggestions** for Hospital Architecture..........................12mo, | 1 | 25 |
| The World's Columbian Exposition of 1893............ 4to, | 2 | 50 |

## ARMY, NAVY, Etc.

MILITARY ENGINEERING—ORDNANCE—PORT CHARGES, ETC.

| | | |
|---|---:|---:|
| Cooke's **Naval Ordnance** ............................8vo, | $12 | 50 |
| Metcalfe's **Ordnance and Gunnery**..........12mo, with Atlas, | 5 | 00 |
| Ingalls's **Handbook of Problems in Direct Fire**...........8vo, | 4 | 00 |
| "       **Ballistic Tables**..........................8vo, | 1 | 50 |
| Bucknill's **Submarine Mines and Torpedoes**..............8vo, | 4 | 00 |
| Todd and Whall's **Practical Seamanship** .................8vo, | 7 | 50 |
| Mahan's **Advanced Guard**............................18mo, | 1 | 50 |
| "       **Permanent Fortifications.** (Mercur.).8vo, half morocco, | 7 | 50 |
| Wheeler's **Siege Operations**...........................8vo, | 2 | 00 |
| Woodhull's **Notes on Military Hygiene**........12mo, morocco, | 2 | 50 |
| Dietz's **Soldier's First Aid**...................12mo, morocco, | 1 | 25 |
| Young's **Simple** Elements of Navigation..12mo, morocco flaps, | 2 | 50 |
| Reed's Signal Service.................................. | | 50 |
| Phelps's Practical **Marine** Surveying......................8vo, | 2 | 50 |
| Very's **Navies of the World**.............8vo, half morocco, | 3 | 50 |
| Bourne's Screw Propellers............................4to, | 5 | 00 |

| | | |
|---|---|---|
| Hunter's Port Charges....................8vo, half morocco, | $13 | 00 |
| *Dredge's Modern French Artillery.........4to, half morocco, | 20 | 00 |
| " Record of the Transportation Exhibits Building, World's Columbian Exposition of 1893..4to, half morocco, | 15 | 00 |
| Mercur's Elements of the Art of War....................8vo, | 4 | 00 |
| " Attack of Fortified Places...................12mo, | 2 | 00 |
| Chase's Screw Propellers..............................8vo, | 3 | 00 |
| Winthrop's Abridgment of Military Law...............12mo, | 2 | 50 |
| De Brack's Cavalry Outpost Duties. (Carr.)....18mo, morocco, | 2 | 00 |
| Cronkhite's Gunnery for Non-com. Officers.....18mo, morocco, | 2 | 00 |
| Dyer's Light Artillery................................12mo, | 3 | 00 |
| Sharpe's Subsisting Armies..........................18mo, | 1 | 25 |
| " " " ..................18mo, morocco, | 1 | 50 |
| Powell's Army Officer's Examiner.....................12mo, | 4 | 00 |
| Hoff's Naval Tactics..................................8vo, | 1 | 50 |
| Bruff's Ordnance and Gunnery.........................8vo, | 6 | 00 |

## ASSAYING.

### SMELTING—ORE DRESSING—ALLOYS, ETC.

| | | |
|---|---|---|
| Furman's Practical Assaying...........................8vo, | 3 | 00 |
| Wilson's Cyanide Processes...........................12mo, | 1 | 50 |
| Fletcher's Quant. Assaying with the Blowpipe..12mo, morocco, | 1 | 50 |
| Ricketts's Assaying and Assay Schemes...............8vo, | 3 | 00 |
| *Mitchell's Practical Assaying. (Crookes.)..............8vo, | 10 | 00 |
| Thurston's Alloys, Brasses, and Bronzes...............8vo, | 2 | 50 |
| Kunhardt's Ore Dressing.............................8vo, | 1 | 50 |
| O'Driscoll's Treatment of Gold Ores..................8vo, | 2 | 00 |

## ASTRONOMY.

### PRACTICAL, THEORETICAL, AND DESCRIPTIVE.

| | | |
|---|---|---|
| Michie and Harlow's Practical Astronomy................8vo, | 3 | 00 |
| White's Theoretical and Descriptive Astronomy..........12mo, | 2 | 00 |
| Doolittle's Practical Astronomy.......................8vo, | 4 | 00 |
| Craig's Azimuth ....................................4to. | 3 | 50 |
| Gore's Elements of Geodesy..........................8vo, | 2 | 50 |

## BOTANY.

### Gardening for Ladies, Etc.

| | | |
|---|---|---|
| Westermaier's General Botany. (Schneider.)............8vo, | $2 | 00 |
| Thomé's Structural Botany.......................18mo, | 2 | 25 |
| Baldwin's **Orchids of New England**................ .....8vo, | 1 | 50 |
| Loudon's **Gardening for** Ladies. (Downing.)...........12mo, | 1 | 50 |

## BRIDGES, ROOFS, Etc.

### Cantilever—Highway—Suspension.

| | | |
|---|---|---|
| Boller's Highway Bridges...................................8vo, | 2 | 00 |
| *  "      The Thames River Bridge.................4to, paper, | 5 | 00 |
| Burr's Stresses in Bridges.... ............................8vo, | 3 | 50 |
| Merriman & Jacoby's **Text-book of Roofs and** Bridges. **Part I., Stresses**..... .... ....................... ........ ...8vo, | 2 | 50 |
| Merriman & Jacoby's **Text-book of** Roofs and Bridges. **Part II., Graphic Statics**................................8vo, | 2 | 50 |
| Merriman & Jacoby's **Text-book of Roofs and Bridges. Part III., Bridge Design**...................................8vo, | 5 | 00 |
| Merriman & Jacoby's **Text-book** of Roofs and Bridges. **Part IV.,** Continuous, Draw, Cantilever, Suspension, and Arched Bridges.......................(*In preparation*). | | |
| Crehore's **Mechanics of the Girder**........................8vo, | 5 | 00 |
| Du Bois's **Strains in Framed Structures**...................4to, | 10 | 00 |
| Greene's **Roof Trusses**.................................8vo, | 1 | 25 |
| "      **Bridge Trusses**...............................8vo, | 2 | 50 |
| "      **Arches in Wood,** etc............................8vo, | 2 | 50 |
| Waddell's **Iron Highway Bridges**.......................8vo, | 4 | 00 |
| Wood's **Construction of Bridges and Roofs**..............8vo, | 2 | 00 |
| Foster's **Wooden Trestle Bridges**........................4to, | 5 | 00 |
| * Morison's **The Memphis Bridge**..................Oblong 4to, | 10 | 00 |
| Johnson's **Modern Framed Structures** ................ ... 4to, | 10 | 00 |

## CHEMISTRY.

### Qualitative—Quantitative—Organic—Inorganic, Etc.

| | | |
|---|---|---|
| Fresenius's Qualitative Chemical Analysis. (Johnson.)....8vo, | 4 | 00 |
| "     Quantitative Chemical Analysis. (Allen.).......8vo, | 6 | 00 |
| "     "     "     "     (Bolton.)......8vo, | 1 | 50 |

| | | |
|---|---|---|
| Crafts's Qualitative Analysis. (Schaeffer.)...........12mo, | $1 | 50 |
| Perkins's Qualitative Analysis..................12mo, | 1 | 00 |
| Thorpe's Quantitative Chemical Analysis............18mo, | 1 | 50 |
| Classen's Analysis by Electrolysis. (Herrick.).........8vo, | 3 | 00 |
| Stockbridge's Rocks and Soils....................8vo, | 2 | 50 |
| O'Brine's Laboratory Guide to Chemical Analysis......8vo, | 2 | 00 |
| Mixter's Elementary Text-book of Chemistry..........12mo, | 1 | 50 |
| Wulling's Inorganic Phar. and Med. Chemistry........12mo, | 2 | 00 |
| Mandel's Bio-chemical Laboratory.................12mo, | 1 | 50 |
| Austen's Notes for Chemical Students..............12mo, | 1 | 50 |
| Schimpf's Volumetric Analysis....................12mo, | 2 | 50 |
| Hammarsten's Physiological Chemistry (Mandel.)......8vo, | 4 | 00 |
| Miller's Chemical Physics........................8vo, | 2 | 00 |
| Pinner's Organic Chemistry. (Austen.)..............12mo, | 1 | 50 |
| Kolbe's Inorganic Chemistry.....................12mo, | 1 | 50 |
| Ricketts and Russell's Notes on Inorganic Chemistry (Non-metallic)....................Oblong 8vo, morocco, | | 75 |
| Drechsel's Chemical Reactions. (Merrill.)............12mo, | 1 | 25 |
| Adriance's Laboratory Calculations................12mo, | 1 | 25 |
| Troilius's Chemistry of Iron......................8vo, | 2 | 00 |
| Allen's Tables for Iron Analysis...................8vo, | 3 | 00 |
| Nichols's Water Supply (Chemical and Sanitary).......8vo, | 2 | 50 |
| Mason's        "          "   ..........................8vo, | 5 | 00 |
| Spencer's Sugar Manufacturer's Handbook.12mo, morocco flaps, | 2 | 00 |
| Wiechmann's Sugar Analysis....................8vo, | 2 | 50 |
| "    Chemical Lecture Notes..................12mo, | 3 | 00 |

## DRAWING.

ELEMENTARY—GEOMETRICAL—TOPOGRAPHICAL.

| | | |
|---|---|---|
| Hill's Shades and Shadows and Perspective............8vo, | 2 | 00 |
| Mahan's Industrial Drawing. (Thompson.)......2 vols., 8vo, | 3 | 50 |
| MacCord's Kinematics.........................8vo, | 5 | 00 |
| "    Mechanical Drawing.....................8vo, | 4 | 00 |
| "    Descriptive Geometry....................8vo, | 3 | 00 |
| Reed's Topographical Drawing. (H. A.)..............4to, | 5 | 00 |
| Smith's Topographical Drawing. (Macmillan.).........8vo, | 2 | 50 |
| Warren's Free-hand Drawing ...................12mo, | 1 | 00 |

| | | |
|---|---|---:|
| Warren's Drafting Instruments....12mo, | | $1 25 |
| " Projection Drawing....12mo, | | 1 50 |
| " Linear Perspective....12mo, | | 1 00 |
| " Plane Problems....12mo, | | 1 25 |
| " Primary Geometry....12mo, | | 75 |
| " **Descriptive** Geometry....2 vols., 8vo, | | 3 50 |
| " Problems and Theorems....8vo, | | 2 50 |
| " Machine Construction....2 vols., **8vo**, | | 7 50 |
| " Stereotomy—Stone **Cutting**....**8vo**, | | 2 50 |
| " Higher Linear Perspective....8vo, | | 3 50 |
| " Shades and Shadows....8vo, | | 3 00 |
| Whelpley's **Letter Engraving**....12mo, | | 2 00 |

## ELECTRICITY AND MAGNETISM.
### Illumination—Batteries—Physics.

| | |
|---|---:|
| * Dredge's Electric Illuminations....2 vols., 4to, half morocco, | 25 00 |
| " " " Vol. II....4to, | 7 50 |
| Niaudet's Electric **Batteries**. (Fishback.)....**12mo**, | 2 50 |
| Anthony and Brackett's Text-book of Physics....8vo, | 4 00 |
| Cosmic Law **of** Thermal Repulsion....18mo, | 75 |
| Thurston's Stationary Steam Engines for Electric Lighting Purposes....12mo, | 1 50 |
| Michie's **Wave Motion Relating to** Sound **and Light,**....**8vo**, | 4 00 |
| Barker's **Deep-sea** Soundings....8vo, | 2 00 |
| Holman's **Precision** of Measurements....8vo, | 2 00 |
| Tillman's **Heat**....8vo, | 1 50 |
| Gilbert's **De-magnete**. (Mottelay.)....**8vo**, | 2 50 |
| Benjamin's Voltaic Cell....8vo, | 3 00 |
| Reagan's Steam and Electrical Locomotives....12mo | 2 00 |

## ENGINEERING.
### Civil—Mechanical—Sanitary, Etc.

| | |
|---|---:|
| * Trautwine's **Cross-section**....Sheet, | 25 |
| * " **Civil Engineer's** Pocket-book...12mo, mor. flaps, | 5 00 |
| * " Excavations and Embankments....8vo, | 2 00 |
| * " Laying Out Curves....12mo, morocco, | 2 50 |
| Hudson's **Excavation Tables.** Vol. II....8vo, | 1 00 |

| | |
|---|---|
| Searles's Field Engineering............12mo, morocco flaps, | $3 00 |
| "   Railroad Spiral ..............12mo, morocco flaps, | 1 50 |
| Godwin's Railroad Engineer's Field-book. 12mo, pocket-bk. form, | 2 50 |
| Butts's Engineer's Field-book...............12mo, morocco, | 2 50 |
| Gore's Elements of Goodesy................8vo, | 2 50 |
| Wellington's Location of Railways.................8vo, | 5 00 |
| *Dredge's Penn. Railroad Construction, etc... Folio, half mor., | 20 00 |
| Smith's Cable Tramways...................4to, | 2 50 |
| "   Wire Manufacture and Uses.............4to, | 3 00 |
| Mahan's Civil Engineering. (Wood.)............8vo, | 5 00 |
| Wheeler's Civil Engineering..................8vo, | 4 00 |
| Mosely's Mechanical Engineering. (Mahan.).........8vo, | 5 00 |
| Johnson's Theory and Practice of Surveying.........8vo, | 4 00 |
| "   Stadia Reduction Diagram..Sheet, $22\frac{1}{2} \times 28\frac{1}{2}$ inches, | 50 |
| *Drinker's Tunnelling...............4to, half morocco, | 25 00 |
| Eissler's Explosives—Nitroglycerine and Dynamite........8vo, | 4 00 |
| Foster's Wooden Trestle Bridges................4to, | 5 00 |
| Ruffner's Non-tidal Rivers....................8vo, | 1 25 |
| Greene's Roof Trusses .....................8vo, | 1 25 |
| "   Bridge Trusses....................8vo, | 2 50 |
| "   Arches in Wood, etc................8vo, | 2 50 |
| Church's Mechanics of Engineering—Solids and Fluids....8vo, | 6 00 |
| "   Notes and Examples in Mechanics..........8vo, | 2 00 |
| Howe's Retaining Walls (New Edition.).............12mo, | 1 25 |
| Wegmann's Construction of Masonry Dams...........4to, | 5 00 |
| Thurston's Materials of Construction..............8vo, | 5 00 |
| Baker's Masonry Construction.................8vo, | 5 00 |
| "   Surveying Instruments...............12mo, | 3 00 |
| Warren's Stereotomy—Stone Cutting.............8vo, | 2 50 |
| Nichols's Water Supply (Chemical and Sanitary).......8vo, | 2 50 |
| Mason's   "   "   "   "   "   .........8vo, | 5 00 |
| Gerhard's Sanitary House Inspection..............16mo, | 1 00 |
| Kirkwood's Lead Pipe for Service Pipe............8vo, | 1 50 |
| Wolff's Windmill as a Prime Mover..............8vo, | 3 00 |
| Howard's Transition Curve Field-book......12mo, morocco flap, | 1 50 |
| Crandall's The Transition Curve............12mo, morocco, | 1 50 |

| | | |
|---|---|---|
| Crandall's Earthwork Tables ...........................8vo, | $1 | 50 |
| Patton's Civil Engineering............................8vo, | 7 | 50 |
| " Foundations.......................8vo, | 5 | 00 |
| Carpenter's Experimental Engineering ................8vo, | 6 | 00 |
| Webb's Engineering Instruments..............12mo, morocco, | 1 | 00 |
| Black's U. S. Public Works............................4to, | 5 | 00 |
| Merriman and Brook's Handbook for Surveyors....12mo, mor., | 2 | 00 |
| Merriman's Retaining Walls and Masonry Dams..........8vo, | 2 | 00 |
| " Geodetic Surveying......................8vo, | 2 | 00 |
| Kiersted's Sewage Disposal..... ......................12mo, | 1 | 25 |
| Siebert and Biggin's Modern Stone Cutting and Masonry...8vo, | 1 | 50 |
| Kent's Mechanical Engineer's Pocket-book.....12mo, morocco, | 5 | 00 |

## HYDRAULICS.

WATER-WHEELS—WINDMILLS—SERVICE PIPE—DRAINAGE, ETC.

| | | |
|---|---|---|
| Weisbach's Hydraulics. (Du Bois.)....................8vo, | 5 | 00 |
| Merriman's Treatise on Hydraulics....................8vo, | 4 | 00 |
| Ganguillet & Kutter's Flow of Water. (Hering & Trautwine.).8vo, | 4 | 00 |
| Nichols's Water Supply (Chemical and Sanitary)..........8vo, | 2 | 50 |
| Wolff's Windmill as a Prime Mover.....................8vo, | 3 | 00 |
| Ferrel's Treatise on the Winds, Cyclones, and Tornadoes...8vo, | 4 | 00 |
| Kirkwood's Lead Pipe for Service Pipe.................8vo, | 1 | 50 |
| Ruffner's Improvement for Non-tidal Rivers............8vo, | 1 | 25 |
| Wilson's Irrigation Engineering.... ....................8vo, | 4 | 00 |
| Bovey's Treatise on Hydraulics.........................8vo, | 4 | 00 |
| Wegmann's Water Supply of the City of New York .......4to, | 10 | 00 |
| Hazen's Filtration of Public Water Supply................8vo, | 2 | 00 |
| Mason's Water Supply—Chemical and Sanitary...........8vo, | 5 | 00 |
| Wood's Theory of Turbines.... ......................8vo, | 2 | 50 |

## MANUFACTURES.

ANILINE—BOILERS—EXPLOSIVES—IRON—SUGAR—WATCHES—
WOOLLENS, ETC.

| | | |
|---|---|---|
| Metcalfe's Cost of Manufactures........................8vo, | 5 | 00 |
| Metcalf's Steel (Manual for Steel Users).................12mo, | 2 | 00 |
| Allen's Tables for Iron Analysis........................8vo, | 3 | 00 |

| | | |
|---|---|---|
| West's American Foundry Practice................12mo, | $2 | 50 |
| "    Moulder's Text-book ........................12mo, | 2 | 50 |
| Spencer's Sugar Manufacturer's Handbook....12mo, mor. flap, | 2 | 00 |
| Wiechmann's Sugar Analysis..........................8vo, | 2 | 50 |
| Beaumont's Woollen and Worsted Manufacture.........12mo, | 1 | 50 |
| * Reisig's Guide to Piece Dyeing......................8vo, | 25 | 00 |
| Eissler's Explosives, Nitroglycerine and Dynamite........8vo, | 4 | 00 |
| Reimann's Aniline Colors. (Crookes.)..................8vo, | 2 | 50 |
| Ford's Boiler Making for Boiler Makers................18mo, | 1 | 00 |
| Thurston's Manual of Steam Boilers...................8vo, | 5 | 00 |
| Booth's Clock and Watch Maker's Manual..............12mo, | 2 | 00 |
| Holly's Saw Filing....................................18mo, | | 75 |
| Svedelius's Handbook for Charcoal Burners............12mo, | 1 | 50 |
| The Lathe and Its Uses................................8vo, | 6 | 00 |
| Woodbury's Fire Protection of Mills....................8vo, | 2 | 50 |
| Bolland's The Iron Founder...........................12mo, | 2 | 50 |
| "    "    "    "   Supplement................12mo, | 2 | 50 |
| "    Encyclopædia of Founding Terms............12mo, | 3 | 00 |
| Bouvier's Handbook on Oil Painting...................12mo, | 2 | 00 |
| Steven's House Painting..............................18mo, | | 75 |

## MATERIALS OF ENGINEERING.

### Strength—Elasticity—Resistance, Etc.

| | | |
|---|---|---|
| Thurston's Materials of Engineering.............3 vols., 8vo, | 8 | 00 |
| Vol. I., Non-metallic............................8vo, | 2 | 00 |
| Vol. II., Iron and Steel..........................8vo, | 3 | 50 |
| Vol. III., Alloys, Brasses, and Bronzes............8vo, | 2 | 50 |
| Thurston's Materials of Construction...................8vo, | 5 | 00 |
| Baker's Masonry Construction.........................8vo, | 5 | 00 |
| Lanza's Applied Mechanics...........................8vo, | 7 | 50 |
| "    Strength of Wooden Columns...........8vo, paper, | | 50 |
| Wood's Resistance of Materials.......................8vo, | 2 | 00 |
| Weyrauch's Strength of Iron and Steel. (Du Bois.).......8vo, | 1 | 50 |
| Burr's Elasticity and Resistance of Materials............8vo, | 5 | 00 |
| Merriman's Mechanics of Materials....................8vo, | 4 | 00 |
| Church's Mechanic's of Engineering—Solids and Fluids.....8vo, | 6 | 00 |

| | | |
|---|---|---|
| Beardslee and Kent's Strength of Wrought Iron.........8vo, | $1 | 5 |
| Hatfield's Transverse Strains............................8vo, | 5 | 00 |
| Du Bois's Strains in Framed Structures..................4to, | 10 | 00 |
| Merrill's Stones for Building and Decoration............8vo, | 5 | 00 |
| Bovey's Strength of Materials...........................8vo, | 7 | 50 |
| Spalding's Roads and Pavements........................12mo, | 2 | 00 |
| Rockwell's Roads and Pavements in France..............12mo, | 1 | 25 |
| Byrne's Highway Construction............................8vo, | 5 | 00 |
| Patton's Treatise on Foundations........................8vo, | 5 | 00 |

## MATHEMATICS.

### CALCULUS—GEOMETRY—TRIGONOMETRY, ETC.

| | | |
|---|---|---|
| Rice and Johnson's Differential Calculus................8vo, | 3 | 50 |
| "    Abridgment of Differential Calculus....8vo, | 1 | 50 |
| "    Differential and Integral Calculus, 2 vols. in 1, 12mo, | 2 | 50 |
| Johnson's Integral Calculus............................12mo, | 1 | 50 |
| "    Curve Tracing..............................12mo, | 1 | 00 |
| "    Differential Equations—Ordinary and Partial.....8vo, | 3 | 50 |
| "    Least Squares..............................12mo, | 1 | 50 |
| Craig's Linear Differential Equations....................8vo, | 5 | 00 |
| Merriman and Woodward's Higher Mathematics..........8vo, | 5 | 00 |
| Bass's Differential Calculus............................12mo, | | |
| Halsted's Synthetic Geometry............................8vo, | 1 | 50 |
| "    Elements of Geometry......................8vo, | 1 | 75 |
| Chapman's Theory of Equations........................12mo, | 1 | 50 |
| Merriman's Method of Least Squares....................8vo, | 2 | 00 |
| Compton's Logarithmic Computations...................12mo, | 1 | 50 |
| Davis's Introduction to the Logic of Algebra..............8vo, | 1 | 50 |
| Warren's Primary Geometry............................12mo, | | 75 |
| "    Plane Problems.............................12mo, | 1 | 25 |
| "    Descriptive Geometry..............2 vols., 8vo, | 3 | 50 |
| "    Problems and Theorems.....................8vo, | 2 | 50 |
| "    Higher Linear Perspective...................8vo, | 3 | 50 |
| "    Free-hand Drawing........................12mo, | 1 | 00 |
| "    Drafting Instruments.......................12mo, | 1 | 25 |

| | | |
|---|---|---|
| Warren's Projection Drawing ............................12mo, | $1 | 50 |
| " Linear Perspective ............................12mo, | 1 | 00 |
| " Plane Problems ..............................12mo, | 1 | 25 |
| Searles's Elements of Geometry. ........................8vo, | 1 | 50 |
| Brigg's Plane Analytical Geometry .....................12mo, | 1 | 00 |
| Wood's Co-ordinate Geometry ...........................8vo, | 2 | 00 |
| " Trigonometry ..............................12mo, | 1 | 00 |
| Mahan's Descriptive Geometry (Stone Cutting) .........8vo, | 1 | 50 |
| Woolf's Descriptive Geometry ....................Royal 8vo, | 3 | 00 |
| Ludlow's Trigonometry with Tables. (Bass.) ...........8vo, | 3 | 00 |
| " Logarithmic and Other Tables. (Bass.) ........8vo, | 2 | 00 |
| Baker's Elliptic Functions ..............................8vo, | 1 | 50 |
| Parker's Quadrature of the Circle .......................8vo, | 2 | 50 |
| Totten's Metrology .....................................8vo, | 2 | 50 |
| Ballard's Pyramid Problem ..............................8vo, | 1 | 50 |
| Barnard's Pyramid Problem .............................8vo, | 1 | 50 |

## MECHANICS—MACHINERY.

### TEXT-BOOKS AND PRACTICAL WORKS.

| | | |
|---|---|---|
| Dana's Elementary Mechanics .........................12mo, | 1 | 50 |
| Wood's " " ..........................12mo, | 1 | 25 |
| " " " Supplement and Key ........... | 1 | 25 |
| " Analytical Mechanics ............................8vo, | 3 | 00 |
| Michie's Analytical Mechanics ............................8vo, | 4 | 00 |
| Merriman's Mechanics of Materials .....................8vo, | 4 | 00 |
| Church's Mechanics of Engineering .....................8vo, | 6 | 00 |
| " Notes and Examples in Mechanics .............8vo, | 2 | 00 |
| Mosely's Mechanical Engineering. (Mahan.) .............8vo, | 5 | 00 |
| Weisbach's Mechanics of Engineering. Vol. III., Part I., Sec. I. (Klein.) ..........................................8vo, | 5 | 00 |
| Weisbach's Mechanics of Engineering. Vol. III., Part I. Sec. II. (Klein.) .........................................8vo, | 5 | 00 |
| Weisbach's Hydraulics and Hydraulic Motors. (Du Bois.) ..8vo, | 5 | 00 |
| " Steam Engines. (Du Bois.) .................8vo, | 5 | 00 |
| Lanza's Applied Mechanics .............................8vo, | 7 | 50 |

| | | |
|---|---|---|
| Crehore's Mechanics of the Girder..........8vo, | $5 | 00 |
| MacCord's Kinematics..........8vo, | 5 | 00 |
| Thurston's Friction and Lost Work..........8vo, | 3 | 00 |
| " The Animal as a Machine..........12mo, | 1 | 00 |
| Hall's Car Lubrication..........12mo, | 1 | 00 |
| Warren's Machine Construction..........2 vols., 8vo, | 7 | 50 |
| Chordal's Letters to Mechanics..........12mo, | 2 | 00 |
| The Lathe and Its Uses..........8vo, | 6 | 00 |
| Cromwell's Toothed Gearing..........12mo, | 1 | 50 |
| " Belts and Pulleys..........12mo, | 1 | 50 |
| Du Bois's Mechanics. Vol. I., Kinematics..........8vo, | 3 | 50 |
| " " Vol. II., Statics..........8vo, | 4 | 00 |
| " " Vol. III., Kinetics..........8vo, | 3 | 50 |
| Dredge's Trans. Exhibits Building, World Exposition, 4to, half morocco, | 15 | 00 |
| Flather's Dynamometers..........12mo, | 2 | 00 |
| " Rope Driving..........12mo, | 2 | 00 |
| Richards's Compressed Air..........12mo, | 1 | 50 |
| Smith's Press-working of Metals..........8vo, | 3 | 00 |
| Holly's Saw Filing..........18mo, | | 75 |
| Fitzgerald's Boston Machinist..........18mo, | 1 | 00 |
| Baldwin's Steam Heating for Buildings..........12mo, | 2 | 50 |
| Metcalfe's Cost of Manufactures..........8vo, | 5 | 00 |
| Benjamin's Wrinkles and Recipes..........12mo, | 2 | 00 |
| Dingey's Machinery Pattern Making..........12mo, | 2 | 00 |

## METALLURGY.

Iron—Gold—Silver—Alloys, Etc.

| | | |
|---|---|---|
| Egleston's Metallurgy of Silver..........8vo, | 7 | 50 |
| " Gold and Mercury..........8vo, | 7 | 50 |
| " Weights and Measures, Tables..........18mo, | | 75 |
| " Catalogue of Minerals..........8vo, | 2 | 50 |
| O'Driscoll's Treatment of Gold Ores..........8vo, | 2 | 00 |
| * Kerl's Metallurgy—Copper and Iron..........8vo, | 15 | 00 |
| * " " Steel, Fuel, etc..........8vo, | 15 | 00 |

| | | |
|---|---|---|
| Thurston's Iron and Steel..............................8vo, | $3 | 50 |
| "    Alloys.......................................8vo, | 2 | 50 |
| Troilius's Chemistry of Iron.............................8vo, | 2 | 00 |
| Kunhardt's Ore Dressing in Europe......................8vo, | 1 | 50 |
| Weyrauch's Strength of Iron and Steel. (Du Bois.)........8vo, | 1 | 50 |
| Beardslee and Kent's Strength of Wrought Iron..........8vo, | 1 | 50 |
| Compton's First Lessons in Metal Working..............12mo, | 1 | 50 |
| West's American Foundry Practice.....................12mo, | 2 | 50 |
| "    Moulder's Text-book...........................12mo, | 2 | 50 |

## MINERALOGY AND MINING.

### Mine Accidents—Ventilation—Ore Dressing, Etc.

| | | |
|---|---|---|
| Dana's Descriptive Mineralogy. (E. S.).....8vo, half morocco, | 12 | 50 |
| "    Mineralogy and Petrography. (J. D.)...........12mo, | 2 | 00 |
| "    Text-book of Mineralogy. (E. S.).................8vo, | 3 | 50 |
| "    Minerals and How to Study Them. (E. S.)......12mo, | 1 | 50 |
| "    American Localities of Minerals..................8vo, | 1 | 00 |
| Brush and Dana's Determinative Mineralogy............8vo, | 3 | 50 |
| Rosenbusch's Microscopical Physiography of Minerals and Rocks. (Iddings.).....................................8vo, | 5 | 00 |
| Hussak's Rock-forming Minerals. (Smith.)..............8vo, | 2 | 00 |
| Williams's Lithology.....................................8vo, | 3 | 00 |
| Chester's Catalogue of Minerals........................8vo, | 1 | 25 |
| "    Dictionary of the Names of Minerals............8vo, | 3 | 00 |
| Egleston's Catalogue of Minerals and Synonyms..........8vo, | 2 | 50 |
| Goodyear's Coal Mines of the Western Coast...........12mo, | 2 | 50 |
| Kunhardt's Ore Dressing in Europe......................8vo, | 1 | 50 |
| Sawyer's Accidents in Mines...........................8vo, | 7 | 00 |
| Wilson's Mine Ventilation............................16mo, | 1 | 25 |
| Boyd's Resources of South Western Virginia.............8vo, | 3 | 00 |
| "    Map of South Western Virginia.....Pocket-book form, | 2 | 00 |
| Stockbridge's Rocks and Soils..........................8vo, | 2 | 50 |
| Eissler's Explosives—Nitroglycerine and Dynamite........8vo, | 4 | 00 |

\*Drinker's Tunnelling, Explosives, Compounds, and Rock Drills.
                                          4to, half morocco, $25 00
Beard's Ventilation of Mines..................... .........12mo,   2 50
Ihlseng's Manual of Mining.. ..........................8vo,   4 00

## STEAM AND ELECTRICAL ENGINES, BOILERS, Etc.
### STATIONARY—MARINE—LOCOMOTIVE—GAS ENGINES, ETC.

Weisbach's Steam Engine. (Du Bois.)...................8vo,   5 00
Thurston's Engine and Boiler Trials.....................8vo,   5 00
   "    Philosophy of the Steam Engine.............12mo,    75
   "    Stationary Steam Engines...................12mo,   1 50
   "    Boiler Explosion.... ......................12mo,   1 50
   "    Steam-boiler Construction and Operation.......8vo,
   "    Reflection on the Motive Power of Heat. (Carnot.)
                                                     12mo,   2 00
Thurston's Manual of the Steam Engine. Part I., Structure
    and Theory.........................................8vo,   7 50
Thurston's Manual of the Steam Engine. Part II., Design,
    Construction, and Operation.......................8vo,   7 50
                                                   2 parts, 12 00
Röntgen's Thermodynamics. (Du Bois.)..................8vo,   5 00
Peabody's Thermodynamics of the Steam Engine......... 8vo,   5 00
   "    Valve Gears for the Steam-Engine.............8vo,   2 50
   "    Tables of Saturated Steam....................8vo,   1 00
Wood's Thermodynamics, Heat Motors, etc...............8vo,   4 00
Pupin and Osterberg's Thermodynamics.................12mo,   1 25
Kneass's Practice and Theory of the Injector ............8vo,   1 50
Reagan's Steam and Electrical Locomotives....... .....12mo,   2 00
Meyer's Modern Locomotive Construction.................4to,  10 00
Whitham's Steam-engine Design ... .....................8vo,   6 00
   "    Constructive Steam Engineering...............8vo,  10 00
Hemenway's Indicator Practice........................12mo,   2 00
Pray's Twenty Years with the Indicator...........Royal 8vo,   2 50
Spangler's Valve Gears................................8vo,   2 50
\* Maw's Marine Engines..................Folio, half morocco,  18 00
Trowbridge's Stationary Steam Engines ...........4to, boards,   2 50

| | | |
|---|---|---|
| Ford's Boiler Making for Boiler Makers..................18mo, | $1 | 00 |
| Wilson's Steam Boilers. (Flather.).....................12mo, | 2 | 50 |
| Baldwin's Steam Heating for Buildings.................12mo, | 2 | 50 |
| Hoadley's Warm-blast Furnace...........................8vo, | 1 | 50 |
| Sinclair's Locomotive Running..........................12mo, | 2 | 00 |
| Clerk's Gas Engine......................................12mo, | 4 | 00 |

## TABLES, WEIGHTS, AND MEASURES.

### For Engineers, Mechanics, Actuaries—Metric Tables, Etc.

| | | |
|---|---|---|
| Crandall's Railway and Earthwork Tables................8vo, | 1 | 50 |
| Johnson's Stadia and Earthwork Tables..................8vo, | 1 | 25 |
| Bixby's Graphical Computing Tables...................Sheet, | | 25 |
| Compton's Logarithms..................................12mo, | 1 | 50 |
| Ludlow's Logarithmic and Other Tables. (Bass.).......12mo, | 2 | 00 |
| Thurston's Conversion Tables............................8vo, | 1 | 00 |
| Egleston's Weights and Measures.......................18mo, | | 75 |
| Totten's Metrology......................................8vo, | 2 | 50 |
| Fisher's Table of Cubic Yards......................Cardboard, | | 25 |
| Hudson's Excavation Tables. Vol. II....................8vo, | 1 | 00 |

## VENTILATION.

### Steam Heating—House Inspection—Mine Ventilation.

| | | |
|---|---|---|
| Beard's Ventilation of Mines ..........................12mo, | 2 | 50 |
| Baldwin's Steam Heating................................12mo, | 2 | 50 |
| Reid's Ventilation of American Dwellings...............12mo, | 1 | 50 |
| Mott's The Air We Breathe, and Ventilation............16mo, | 1 | 00 |
| Gerhard's Sanitary House Inspection...........Square 16mo, | 1 | 00 |
| Wilson's Mine Ventilation..............................16mo, | 1 | 25 |
| Carpenter's Heating and Ventilating of Buildings........8vo, | 3 | 00 |

## MISCELLANEOUS PUBLICATIONS.

| | | |
|---|---|---|
| Alcott's Gems, Sentiment, Language..............Gilt edges, | 5 | 00 |
| Bailey's The New Tale of a Tub..........................8vo, | | 75 |
| Ballard's Solution of the Pyramid Problem..............8vo, | 1 | 50 |
| Barnard's The Metrological System of the Great Pyramid..8vo, | 1 | 50 |

| | |
|---|---|
| * Wiley's Yosemite, Alaska, and Yellowstone ............4to, | $3 00 |
| Emmon's Geological Guide-book of the Rocky Mountains..8vo, | 1 50 |
| Ferrel's Treatise on the Winds............................8vo, | 4 00 |
| Perkins's Cornell University................Oblong 4to, | 1 50 |
| Ricketts's History of Rensselaer Polytechnic Institute.....8vo, | 3 00 |
| Mott's The Fallacy of the Present Theory of Sound .Sq. 16mo, | 1 00 |
| Rotherham's The New Testament Critically Emphathized. 12mo, | 1 50 |
| Totten's An Important Question in Metrology. ...........8vo, | 2 50 |
| Whitehouse's Lake Mœris............................Paper, | 25 |

## HEBREW AND CHALDEE TEXT-BOOKS.

FOR SCHOOLS AND THEOLOGICAL SEMINARIES.

| | |
|---|---|
| Gesenius's Hebrew and Chaldee Lexicon to Old Testament. (Tregelles.)............ .............. Small 4to, half morocco, | 5 00 |
| Green's Grammar of the Hebrew Language (New Edition).8vo, | 3 00 |
| " Elementary Hebrew Grammar... ..............12mo, | 1 25 |
| " Hebrew Chrestomathy..........................8vo, | 2 00 |
| Letteris's Hebrew Bible (Massoretic Notes in English). 8vo, arabesque, | 2 25 |
| Luzzato's Grammar of the Biblical Chaldaic Language and the Talmud Babli Idioms..............................12mo, | 1 50 |

## MEDICAL.

| | |
|---|---|
| Bull's Maternal Management in Health and Disease.......12mo, | 1 00 |
| Mott's Composition, Digestibility, and Nutritive Value of Food. Large mounted chart, | 1 25 |
| Steel's Treatise on the Diseases of the Ox.... ............8vo, | 6 00 |
| " Treatise on the Diseases of the Dog...............8vo, | 3 50 |
| Worcester's Small Hospitals—Establishment and Maintenance, including Atkinson's Suggestions for Hospital Architecture..... .............................. ............12mo, | 1 25 |
| Hammarsten: Physiological Chemistry. (Mandel.)........8vo, | 4 00 |

TE
How

122023

Author Howe; M.A.

Title Retaining Walls for Earth

www.ingramcontent.com/pod-product-compliance
Lightning Source LLC
Chambersburg PA
CBHW030258170426
43202CB00009B/792